T0200742

ENGINEERING COMPLEX PHENOTYPES IN INDUSTRIAL STRAINS

ENGINEERING COMPLEX PHENOTYPES IN INDUSTRIAL STRAINS

Edited by

RANJAN PATNAIK

AIChE

A JOHN WILEY & SONS, INC., PUBLICATION

Library of Congress Cataloging-in-Publication Data:

Engineering complex phenotypes in industrial strains / edited by Ranjan Patnaik.
    p. cm.
  Includes bibliographical references and index.
    ISBN 978-0-470-61075-6 (hardback)
  1. Industrial microorganisms. 2. Genetic engineering. I. Patnaik, Ranjan, 1969–
QR53.E54 2012
579'.163–dc23

                                                                    2012015254

Printed in the United States of America

10 9 8 7 6 5 4 3 2 1

CONTENTS

FOREWORD

The increasing demands for renewable chemicals, materials, and fuels, as well as the continuing evolution of capabilities in biology, chemistry, and engineering, are giving rise to significant efforts in using biotechnological approaches in new process configurations. These approaches are particularly well suited to conversions of carbohydrate and other biological starting compounds into useful materials, as enzymes and microbes naturally transform these substances. Building on a fair history of industrial use of microbes in the production of high-value, low-volume materials, such as pharmacologically active compounds, vitamins, and amino acids, we are now extending these approaches to the production of higher volume/lower value chemicals, such as monomers for making polymers, lubricants, and fuels. As we progress up this volume curve, the demands on the bioprocess become more and more stringent, and highly integrative approaches among disciplines are required to produce the biocatalysts and associated processes necessary for commercially viable outcomes.

Coincident with this evolution, a number of books and monographs have appeared on the subject of metabolic engineering and systems biology, and the primary literature is becoming more and more detailed. With this backdrop, this book does not attempt to be an authoritative reference on tools and techniques, but rather focuses on the strategies and approaches that enable commercial biocatalyst design. It should be of use to graduate students and early career professionals in the field, or to other generalists and professionals from related disciplines who are eager to grasp the basic tenets of engineering biocatalysts. In addition, it may well be of value in providing corporate managers and government officials with insights into the requirements for successful program outcomes.

This book gives an overview of current approaches, with examples drawn from academia and industry and covering biocatalysts ranging from *Escherichia coli* and *Steptomyces* to yeast and microalgae. The vitality of the field is exemplified by the relatively young ages of the contributors, who are shaping the field with their novel approaches, and the inclusion of case studies adds a realistic dimension to the exposition.

JOHN PIERCE

London

PREFACE

This book highlights current trends and developments in the area of engineering industrial strains for the production of bulk chemicals and biofuels from renewable biomass. The commercialization of bioprocesses derived from the use of superior engineered strains often requires the balance between unknowns and trade-off between multiple complex traits of the biocatalyst. Complex phenotypes are traits in a microbe that require more than one genetic change (multigenic) to be modulated simultaneously in the microorganism's genome for full expression. Knowing what those genetic changes are for a given trait and how to manipulate those targets in the most efficient way, forms the motivation for writing this book. The chapters address tools and methodologies developed for engineering such complex traits or phenotypes in industrial strains. Emphasis is on the multidisciplinary (metabolic engineering, screening, fermentation, downstream) nature of the approach or strategy that is used during the course of developing such a commercial biocatalyst. Keeping in perspective the multidisciplinary nature of activity and the interests of a broader range of readers, the topics included in the chapters are not meant to be fully exhaustive in their respective areas; rather, the emphasis is on comparison and integration of different tools and objectives. Chapters 1–5 summarize broadly the current tools and technologies available for engineering a complex phenotype in an industrial strain with brief reference to examples, while Chapters 6–9 highlight in detail the application of such tools and methodologies in the form of case studies.

Chapter 1 summarizes the age-old proven approach for engineering industrial strains using mutagenesis, followed by screening or selection, often termed classical strain engineering (CSI). Discussions of the applicability of CSI for engineering complex traits provide information on its suitability and limitations. Chapter 2 describes the current state of the art in the use of ^{13}C tracer-based analysis and metabolic flux analysis for engineering complex pathways. Chapter 3 describes the utility of genome-scale models by integration of "omics" technology and physiological data to address engineering of complex traits.

The probability of commercial success of a bioprocess that uses microbial catalysts and renewable feedstocks, as compared with platforms that use chemical catalysts and fossil fuel-derived feedstocks, greatly depends on the time it takes to engineer these microbes to perform the desired reaction under harsh manufacturing conditions at rates, titers, and yields that meet the criteria for economic feasibility. Chapter 4 addresses new evolutionary strain engineering approaches that are superior to CSI in developing complex traits rapidly. Transitioning from laboratory-scale demonstration to commercial-scale operation is not only time-consuming but also expensive, especially with the uncertainties associated with scalability of complex traits. Chapter 5 describes an integrative platform for rapid fermentation process development and strain evaluation that not only minimizes the number of false positives from a strain engineering program but also provides a cost-effective approach to optimize fermentation conditions.

Chapter 6 is a case study on the use of CSI (Chapter 1) and improved strain screening strategies (Chapter 5) at Dutch State Mines for engineering *Streptomyces clavuligerus* for commercial production of anti-infectives. Chapter 7 is a case study on the use of evolutionary approaches (Chapter 4) at Opx Biotechnologies for improving tolerance of *Escherichia coli* to 3-hydroxypropionoic acid. Chapter 8 is a complete strain engineering case study from the National Renewable Energy Laboratory in an unsequenced microalga, *Chlorella vulgaris*, for production of biofuels. The authors have highlighted integration of improved analytics and strain screening approaches (Chapters 1 and 5) with "omics" technology (Chapter 3) for addressing needed improvements in multiple complex traits. Chapter 9 demonstrates the feasibility of using genome-shuffling approaches (Chapter 4) in *Saccharomyces cerevisiae* and *Schefferomyces stiptis* for improving tolerance to inhibitors in lignocellulosic substrates.

Scientists, engineers, and project managers who are leaders in their respective areas of research and drawn from diverse fields of science and engineering have contributed to the above chapters. The book has attempted to capture the thought processes on which they so often rely during the initiation and development of a commercial biocatalyst project. I hope the readers find the content of the book to be intellectually satisfying.

I would like to thank the editors at John Wiley & Sons for being patient and for their cooperation during the course of this project.

RANJAN PATNAIK

CONTRIBUTORS

Hal S. Alper, Ph.D., Department of Chemical Engineering, University of Texas, Austin, TX

Lawrence Chew, Ph.D., Pfenex Inc., San Diego, CA

Yat-Chen Chou, National Renewable Energy Laboratory, Golden, CO

Nathan Crook, Department of Chemical Engineering, University of Texas, Austin, TX

Michael Dauner, Ph.D., E. I. du Pont de Nemours and Company, Wilmington, DE

Bryon S. Donohoe, Ph.D., National Renewable Energy Laboratory, Golden, CO

Jing Du, Ph.D., Department of Chemical and Biomolecular Engineering, University of Illinois at Urbana-Champaign, Urbana, IL

Ryan T. Gill, Ph.D., Department of Chemical and Biological Engineering, University of Colorado-Boulder, Boulder, CO

Michael T. Guarnieri, Ph.D., National Renewable Energy Laboratory, Golden, CO

Marcus Hans, Ph.D., DSM Biotechnology Center, Delft, The Netherlands

Byoungjin Kim, Ph.D., Energy Biosciences Institute, University of Illinois at Urbana-Champaign, Urbana, IL

Eric P. Knoshaug, National Renewable Energy Laboratory, Golden, CO

Bert Koekman, Ph.D., DSM Biotechnology Center, Delft, The Netherlands

Lieve M.L. Laurens, Ph.D., National Renewable Energy Laboratory, Golden, CO

Matthew L. Lipscomb, Ph.D., OPX Biotechnologies, Inc., Boulder, CO

Tanya Warnecke Lipscomb, Ph.D., OPX Biotechnologies, Inc., Boulder, CO

Michael D. Lynch, Ph.D., OPX Biotechnologies, Inc., Boulder, CO

Vincent J.J. Martin, Ph.D., Department of Biology, Centre for Structural and Functional Genomics, Concordia University, Montréal, Québec, Canada

Robin Osterhout, Ph.D., Genomatica, Inc., San Diego, CA

Ranjan Patnaik, Ph.D., Head Biofuels R&D, DuPont India Private Ltd., Hyderabad, India

Priti Pharkya, Ph.D., Genomatica, Inc., San Diego, CA

Philip T. Pienkos, Ph.D., National Renewable Energy Laboratory, Golden, CO

John Pierce, Ph.D., Chief Bioscientist, BP, London, UK

Dominic Pinel, Department of Biology, Centre for Structural and Functional Genomics, Concordia University, Montréal, Québec, Canada

Jun Sun, Ph.D., E. I. du Pont de Nemours and Company, Newark, DE

Stephen Van Dien, Ph.D., Genomatica, Inc., San Diego, CA

Huimin Zhao, Ph.D., Department of Chemical and Biomolecular Engineering, University of Illinois at Urbana-Champaign, Urbana, IL

1

CLASSICAL STRAIN IMPROVEMENT

Nathan Crook and Hal S. Alper

1.0 INTRODUCTION

Improving complex phenotypes, which are typically multigenic in nature, has been a long-standing goal of the food and biotechnology industry well before the advent of recombinant DNA technology and the genomics revolution. For thousands of years, humans have (whether intentionally or not) placed selective pressure on plants, animals, and microorganisms, resulting in improvements to desired phenotypes. Clear evidence of these efforts can be seen from the dramatic morphological changes to food crops since domestication (1). These improvements have been predominantly achieved through a "classical" approach to strain engineering, whereby phenotypic improvements are made by screening and mutagenesis of strains that use methods naive of genome sequences or the resulting genetic changes. This approach is well suited for strain optimization in industrial microbiology, which commonly exploits complex phenotypes in organisms with poorly defined or monitored genetics. As a recognition of importance, Arnold Demain and Julian Davies begin their *Handbook of Industrial Microbiology and Biotechnology* with "Almost all industrial microbiology processes require the initial isolation of cultures from nature, followed by small-scale cultivations and optimization, before large-scale production can become a reality" (2). The classical approach is concerned

Engineering Complex Phenotypes in Industrial Strains, First Edition. Edited by Ranjan Patnaik.
© 2013 John Wiley & Sons, Inc. Published 2013 by John Wiley & Sons, Inc.

with the central steps in this process—between isolation and large-scale production. Hence, the methods and techniques utilized in this approach amount
to "unit operations," that is, standard procedures that can be generically
applied to any desired strain of interest.

A variety of approaches are used to force genetic (and hence phenotypic)
diversity including naturally occurring genetic variation and genetic drift,
mutagenesis, mating/sporulation, and/or selective pressures. These methods
have garnered large successes across a wide range of host organisms owing
mostly to the absence of required sophisticated genomic information or
genetic tools (3). Thus, the classical approach can be applied to both model
organisms (such as *Escherichia coli* and *Saccharomyces cerevisiae*) and newly
isolated or adapted industrial strains. As a result, the classical approach has
seen wide adoption in industrial fermentations due to its proven track record
in alcohol and pharmaceutical production. Finally, strains developed in this
manner are currently accorded non-genetically modified organism (GMO)
status, removing significant barriers to their acceptance by both regulatory
agencies and consumers. This chapter will highlight several of the approaches
and successes that exemplify the classical approach for improving complex
phenotypes of industrial cells as well as indicate its limitations and potential
interfaces with emerging technology.

1.1 THE APPROACH DEFINED

The classical approach is characterized by the introduction of random mutations (either forced or natural) to a population of cells followed by screening
and/or selection to isolate improved variants. The defining quality of classical
strain engineering (as opposed to other evolutionary engineering methods) is
genome-wide mutagenesis. This approach utilizes techniques that introduce
variation across all regions of the genome, in contrast to other techniques that
specifically target the mutations to single genes (or subsequences thereof). To
date, this approach has been successful in improving complex phenotypes
because of the global nature of classical methodologies (see Box 1.1 in this
chapter and case study in Chapter 6). Complex phenotypes such as tolerance
to environmental stress, altered morphology, and improved flocculation characteristics are often influenced by the interactions between multiple (often
uncharacterized) genes. In contrast, without significant prior understanding,
variants generated through mutagenesis of specific genomic subsections are
unlikely to gain proper coverage of the genotype. Indeed, as will be discussed
later, this approach has continuously yielded improved variants for a wide
variety of complex biotechnological applications. The theory and techniques
for the two major steps of classical strain improvement (CSI) (mutagenesis
and screening) are the focus of this chapter, including practical recommenda-
tions for their implementation as well as brief discussion of examples of each
method's industrial application.

BOX 1.1: APPLICATION OF CSI IN SAKE FERMENTATION

The Japanese-brewed sake is produced from rice mash using *Aspergillus oryzae* to saccharify the rice and strains of sake yeast (genus *Saccharomyces cerevisiae*) to ferment the sugars to ethanol. The ideal process imposes a number of complex traits on the sake yeast, including high fermentation capacity over the 20- to 25-day process at low temperatures (typically 10°C), high ethanol tolerance (ethanol levels can approach 15–20%), minimal foaming, resistance to contaminating microbes, and the ability to create the correct proportion of flavor components including higher alcohols and esters (82). Many of these traits have been approached using methods of the classical approach including mutagenesis, selection, and cell mating. Specifically, UV and chemical mutagenesis have dominated as a means of retaining GRAS status for this yeast. Moreover, difficulty in sporulation has limited genetic dissection and a more rational approach until recently (83). Natural selection and isolation from hundreds of years of fermentation has resulted in the series of commonly used strains named the Kyokai series, with Kyokai no. 7 and Kyokai no. 9 as the main fermentation strains used industrially. Due to the superior brewing capacity of Kyokai no. 7, many attempts have been made to improve this strain through the classical approach as well as dissect the underlying genetic changes. Recently, it has been demonstrated that the breeding and selection process of this strain resulted in heterozygosity of many alleles responsible for ethanol production and aromatic compound synthesis (84,85) as seen by sporulation analysis. Many attempts have been made to improve the characteristics of Kyokai no. 7. Non-foaming mutants have been isolated from spontaneous clones as well as UV-induced mutants using selection methods such as cell agglutination and froth floatation (86). Improved strains have also been isolated through chemical mutagenesis (e.g., by EMS) to select for improved flavor profiles. In this case, mutant Kyokai no. 7 strains more resistant to cerulenin were thought to produce more ethyl caproate, an important flavor component. This approach was successful in improving this flavor component; however, the complete portfolio of complex phenotypes was not fully assayed (47). Finally, prevention of contaminants has been explored through mating sake yeast strains with strains exhibiting the killer phenotype (56), which would ward off contaminating yeasts. Collectively, these examples of complex phenotype engineering highlight the difficulties of the process, specifically; it is often hard to create all traits at once. The evolution of the sake yeast demonstrates the power of the classical approach. More recent attempts have been made to use the rational or evolutionary approach for this strain; however, Kyokai no. 7 remains the industrial favorite for sake production.

1.2 MUTAGENESIS

A fundamental parameter dictating success in classical strain engineering is the frequency and type of mutation applied to the parent cells. Typically, this rate is determined by the dose and type of mutagen delivered. To test mutagen specificity and rate, it is common to generate an inactive (mutant) form of some easily assayable gene (e.g., *LacZ* in *E. coli*) that differs from the wild-type gene by a single base-pair change, and test the frequency of reversion. For example, Cupples et al. generated six variants of *LacZ* to show that many common mutagens (EMS, NTG, 2-aminopurine, and 5-azacytidine) are in fact quite specific for certain base-pair changes in *E. coli* (4). Hampsey undertook a similar approach in *S. cerevisiae* and found similarly that mutagens were highly specific. However, the mutation frequencies and specificities were significantly different from those observed in *E. coli* (5). Frameshift and deletion frequencies can also be detected through analysis of a cleverly mutated marker (6). Through analyses of reversion frequencies or high-throughput sequencing, a detailed picture of a treatment's mutagenic profile may be ascertained. This detailed information can be then be used to compute several useful quantities, such as the average number of mutations per genome or the expected number of distinct variants among a mutated population. Knowledge of these frequencies and landscapes are especially useful when designing a selection program, as detection of rare variants (e.g., individuals possessing certain particular mutations and no more) will require many individuals to be screened, whereas more probable patterns of mutagenesis (e.g., if additional silent or neutral mutations are tolerable) will not. At the same time, more focused patterns of mutation inherently limit the search space.

1.2.1 Numerical Considerations in Screen Design

Although in general every possible base substitution will occur at a different frequency (and vary nonuniformly throughout the genome), it is instructive to neglect deletions or insertions and assume all base changes at each site are equiprobable (i.e., occur at the same frequency) to make use of the binomial distribution, to obtain approximate probabilities of any desired mutagenic outcome. If the probability of a single base being mutated to any other base is p, then the probability that a genome of size g has n mutations after mutagenesis is:

$$P(g, n, p) = \frac{g!}{n!(g-n)!} p^n (1-p)^{g-n}.$$

By using well-known properties of the binomial distribution, the average number of mutations per genome is gp with variance $gp(1-p)$. Random genetic drift results in mutation rates of 10^{-10} to 10^{-5}, while forced mutagenesis can elicit rates upwards of 10^{-3} as described below, so this will restrict the range of p. It is apparent that if p is too low (that is, less than $1/g$), there will be many variants with few or no mutations and a vanishingly small population of highly mutated individuals. Furthermore, the binomial coefficient indicates that libraries with low mutation rate (and thus a high population of slightly mutated individuals) are very likely to be redundant, that is, have many individuals of the same genotype. Thus, it is of interest to know the expected number of *distinct* variants in a mutant library to guide screen design. Patrick et al. developed a suite of algorithms to compute many quantities of interest for screening a mutant pool derived from a mutagenic procedure of arbitrary specificity, including the expected number of distinct mutants following mutagenesis (7,8). If the library is highly redundant, then screening of the entire mutated population may not be necessary to ensure complete coverage. As diversity increases, however, the required screening fraction will approach unity. Since complex phenotypes are controlled by the action of multiple genes, high mutation rates are often employed, generally resulting in high library diversity and a strong incentive to screen the entire mutated pool.

To choose the correct rate of mutagenesis and screening, it is important to know the rarity of the phenotype of interest. In the worst and most restrictive case, an improved phenotype will be acquired by mutants containing only a certain set of mutations. For example, consider a particular phenotype that only manifests itself when n-specific mutations are present and no more. In this case, one must determine the mutation rate $\hat{p}$ which maximizes the fraction of n-mutant variants in the mutated population (using one of the tools mentioned earlier) and screen until a reasonably high probability of complete coverage is achieved. For a genome of g base pairs, we can take the derivative of the binomial distribution with respect to mutation rate and set it equal to zero:

$$\frac{d}{dp}\left(\frac{g!}{n!(g-n)!}p^n(1-p)^{g-n}\right)=0.$$

Eliminating constants and taking the derivative, we have:

$$\hat{p}^{n-1}(1-\hat{p})^{g-n-1}(n-g\hat{p})=0.$$

The obvious interesting candidate for a solution is:

$$\hat{p}=\frac{n}{g}.$$

Taking the second derivative of the binomial distribution yields:

$$\frac{g!}{n!(g-n)!}\hat{p}^{n-2}(1-\hat{p})^{g-n-2}\left(n(-2(g-1)\hat{p}-1)+(g-1)g\hat{p}^2+n^2\right).$$

Because

$$\frac{g!}{n!(g-n)!}\hat{p}^{n-2}(1-\hat{p})^{g-n-2}>0,$$

we can substitute our candidate solution into the remaining portion of the second derivative to determine its sign:

$$\left(n\left(-2(g-1)\frac{n}{g}-1\right)+(g-1)g\left(\frac{n}{g}\right)^2+n^2\right)=\frac{n^2}{g}-n,$$

which is clearly negative for $g > n$. Hence, the likelihood of attaining n mutations in a genome of size g is maximized when the mutation rate is n/g. This maximum likelihood is:

$$r\equiv\frac{g!}{n!(g-n)!}\left(\frac{n}{g}\right)^n\left(1-\frac{n}{g}\right)^{g-n}.$$

It is generally necessary to screen more than the number of possible mutants to ensure coverage of the diversity. To obtain, on average, F fractional coverage of all n-mutant variants, it will be necessary to solve

$$F=1-e^{-r*a*L}$$

for L, where a is the probability of selecting the correct n-mutant variant ($1/V$ in this case, where V is the number of possible n-mutant variants [given by the binomial coefficient]) and L is the library size (7). For a small-sized genome (10^6 base pairs) and a phenotype requiring two specific mutations (hence at an optimal mutation rate of $2*10^{-6}$), L works out to be $5.5*10^{12}$ to obtain 95% coverage, on average, which is outside the scope of most screening programs (Assuming a standard yeast cell density of 10^7 per mL and an average cell sorting rate of 10^3 per second, screening this many individuals would require 550 L of culture (for growth-based selections) or 241 years of cell sorting [for fluorescence-based screens]!)

Luckily, most complex phenotypes can tolerate the existence of additional silent/neutral mutations. To account for a small number of allowable neutral mutations, let us assume that the desired n mutations may be found in any variant containing up to $m > n$ total mutations, but no more. The analysis for this case proceeds in much the same way as before with one minor alteration resulting from the fact that a variant with $m > n$ mutations contains

$$\frac{m!}{n!(m-n)!}$$

instances of n mutations. Therefore, maximizing the quantity

$$\sum_{i=n}^{m} \frac{i!}{n!(i-n)!} * P(g,i,p)$$

with respect to p will yield the mutation rate $\hat{p}$ which maximizes the number of n-mutant combinations encountered in the randomized pool. This rate may be used to find the probability that a variant selected at random will have between n and m mutations:

$$r \equiv \sum_{i=n}^{m} P(g,i,\hat{p}).$$

Given that a variant has between n and m mutations, the probability that it contains the mutations of interest can be obtained by summing the probabilities of finding the mutations of interest at each particular mutational level:

$$a = \sum_{i=n}^{m} \frac{P(g,i,\hat{p})}{\sum_{i=n}^{m} P(g,i,\hat{p})} * \frac{i!}{n!(i-n)!} * \frac{1}{V}$$

where

$$\frac{P(g,i,\hat{p})}{\sum_{i=n}^{m} P(g,i,\hat{p})}$$

is the probability that a variant with a mutation rate i is selected, and

$$\frac{i!}{n!(i-n)!} * \frac{1}{V}$$

is the probability of finding a particular combination of n mutations within that variant. r and a can then be substituted to the equation for F, which is solved for L as before. Continuing with the example stated above, if the search is expanded to allow desired mutations to occur in a background of up to 5 mutations, then the mutation rate can be increased to $4.2*10^{-6}$, requiring screening of $4.3*10^{11}$ individuals, which, though an order of magnitude less than in the previous case, is still rather unmanageable.

The property that allows strain engineering programs to be feasible is the additivity of the effects of mutations; that is, even if a particular combination of 12 point mutations is optimal, a couple of them, even in isolation, will be beneficial. This allows engineering to proceed in several single mutation steps as opposed to a single multiple-mutation bound. Because any given single mutation is much more probable than a particular double mutation, the probability of isolating improved variants is greatly increased. Even if a phenotype could only be improved by a single base-pair change in the absence of any others, the number of mutants that must be screened is $8.2*10^6$, which is attainable from a fraction of a milliliter of culture under growth selection or from less than an hour of cell sorting (9).

These probabilities guide strain selection. In the following sections, multiple mutagenesis techniques and screening strategies will be discussed. Particular attention should be paid to mutation rate and throughput, respectively, so that screening programs are designed and carried out efficiently.

1.2.2 Random Genetic Drift

Natural mutations due to errors in replication take place at frequencies between 10^{-5} and 10^{-10} , depending on the strain and organism. Given this frequency, it is not surprising that large-scale selections are required to isolate any improved mutant. These frequencies are supported by a meta-analysis of phenotype occurrence frequencies for the basic yeast *S. cerevisiae* (10). Phenotype reversions requiring single base-pair changes including amino acid auxotrophy reversion and resistances occurred, on average, at a frequency of 10^{-8}. On the basis of this low mutation rate alone, it would appear that random genetic drift may be most suited for the optimization of phenotypes under the control of nonepistatic factors, despite the prevalence of epistatic interactions in nature (11). However, single point mutations are not the only genetic change to take place in evolving cells. Specifically, Lenski et al. found that the majority of the genotypic changes observed through the course of a long-term natural evolution experiment in *E. coli* resulted from transpositions and rearrangements as opposed to single base-pair substitutions (12). These large-scale genetic changes have a much higher probability of generating mutants on distant peaks than do point mutations. The variety of possible genetic changes resulting from a natural evolution program points to its versatility in optimizing a wide variety of complex phenotypes. However, generation of mutants with this method requires a significant amount of time, during which individuals are subjected to growth-based selective pressures. If the phenotype of interest is at odds with growth, then this mutagenic procedure may not be optimal with respect to library size and screening. Natural mutagenesis, however, does lend itself very well to growth phenotypes, as no additional effort on the part of the strain engineer is required to generate mutants and compare them against the fittest variant.

1.2.2.1 Tracking Evolution through Neutral Phenotypes Since natural evolution experiments are often accomplished in continuous liquid culture, new variants are constantly being generated and compared against the fittest variant. Therefore, evolution does not proceed in rounds or stages like most forced evolution experiments, and so it is unclear when fitness increase has ceased or when a population has stopped evolving. To overcome this limitation, one can make use of neutral markers to detect a mutation event. Neutral markers are genotypes that confer no alteration in growth rate yet whose

phenotype is easily detectable. Common neutral phenotypes include resistance
to phage T5 or reversion to lactose fermentation (when the appropriate selective pressure is not present in the screening medium, of course). In a strain
normally deficient in either of these phenotypes, it is expected that over the
course of evolution these neutral mutations will become fixed in the evolving
population at a slow but steady rate. Due to the low probability of attaining
the neutral marker and the fact that it does not pose a selective pressure on
those who carry it, it is assumed that neutral markers are never predominant
in the culture and that over time the prevalence of this phenotype will attain
a steady state as the rate of incidence becomes balanced by mutational losses.
When a variant of high competitive fitness is generated (which in all likelihood
does not possess the neutral marker), it steadily outcompetes the existing
population, driving the proportion of the neutral marker down. After this
new variant becomes predominant in the culture, the fraction of mutants containing the neutral marker again increases at a slow but steady rate as before.
As a result, these sharp declines in the presence of a neutral marker signal the
appearance of a new adaptive mutation, and the jagged graph of the neutral
marker over time is called a periodic selection curve (see Figure 1.1). However,
if the adaptive mutation happens to occur on a variant containing the neutral
marker, the fraction of the neutral marker will approach unity in the selective
medium, although it confers no selective advantage. This phenomenon is
known as "hitchhiking," and it is predicted to occur often in evolving systems
(13,14). In addition, the appearance of an additional adaptive mutation does
not necessarily imply any significant improvement in phenotype as desired by
the strain engineer, as fixation of mutations may simply result in a competitive
advantage quite unrelated to the phenotype of interest. Therefore, it is necessary to assay for improvement as selection proceeds, to ensure that progress
is still occurring.

1.2.2.2 Genetic Determinants of Mutation Rate   Although a wide range
of chemical and physical agents can serve to mutate a microbial population
(discussed later), a variety of more natural factors can contribute to an increase
in the mutation rate of naturally evolving populations, including ploidy, genotype, and environmental conditions. Diploid strains have been shown to more
quickly fix adaptive mutations than haploids. Since diploid strains make copies
of genes at twice the rate of haploid strains, adaptive mutations are generated
at twice the rate. Assuming adaptive mutations are dominant, this increased
generation of adaptive mutations should result in an increase in the rate of
fixation of adaptive mutations. Indeed, Paquin and Adams showed that diploid
strains accumulated advantageous mutations at 1.6 times the rate of haploid
strains (15). This implies that diploid strains may achieve maximum phenotypic increase much faster than haploid strains, allowing further selection
programs to be undertaken.

   In addition to diploid strains, a number of additional "mutator" genotypes
are known to increase the mutation rate in bacteria (16,17). These genotypes

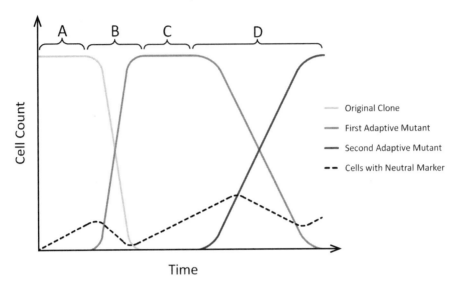

FIGURE 1.1. Periodic selection in an asexual population. The numbers of successive adaptive clones and the number of clones possessing the neutral marker are tracked over time. (A) The prevalence of the neutral marker increases in the wild-type population at a rate determined by natural mutation frequency. (B) The first adaptive mutant (with a large selective advantage) appears in the neutral marker-null population and quickly outcompetes the original clone, causing a sharp decline in the number of cells containing the neutral marker. (C) The prevalence of the neutral marker increases in the first adaptive mutant at a rate determined by natural mutation frequency. (D) The second adaptive mutant (with a relatively small selective advantage) appears in the neutral marker-null population and gradually outcompetes the original clone, causing a slower decline in the number of cells containing the neutral marker than in time interval (B). Reproduced with permission of Annual Reviews, Inc., from Reference (14); permission conveyed through Copyright Clearance Center, Inc.

may encode for enzymes that are naturally mutagenic or may confer mutagenic activity upon an existing enzyme. Although the presence of a gene conferring a high mutation rate would appear detrimental, it has been theorized that a gene conferring a 1000-fold increase in mutation rate in a particular individual can cause a population to increase in fitness quite quickly while remaining in but a small fraction of individuals (18). The ability of this genotype to confer a selective advantage without becoming ubiquitous can be understood through the high rate of reversion of the mutator genotype, due to its high mutation rate. This implies that mutator strains may be an excellent starting point for a variety of evolution experiments, as isolates from the resulting culture are likely to be genetically stable. However, it should be noted that most known mutator genes only achieve a 100-fold increase in mutation frequency, at which level they have been theorized to attain a much larger fraction of the population (5–10%) (16,17). In addition, the mutator pheno-type may be amplified by certain chemicals such as thymidine, allowing for increased control over evolutionary rate over the course of the experiment. Mutator genes are also unique in that they may have a very specific mutational spectrum. In *E. coli*, mutY increases the frequency of GC->TA transversions, mutT results in TA->GC transversions, and the mutD5 mutation appears to increase the rate of mutation in a nonspecific manner. The specificity of mutator genotypes allows a great deal of control over the spectrum of mutants generated, possibly enabling preservation of a counter-selected genotype that is necessary for the application of interest.

1.2.2.3 *Applications of Random Genetic Drift*

Continuous culturing and serial transfers have been successfully used to select for fast-growing strains generated through a natural evolution program (19–22). Included in this list are improvements of basal-level growth rate as well as improvements in growth rate on alternative sugars such as xylose. Accumulated mutations in a yeast strain selected on xylose over time resulted in greatly altered xylose transport kinetics, doubling V_{max} (15.8 to 32 mmol/[g dry weight]/h) and reducing K_m by 25% (132 to 99 mM) (22). A second highlighted example involves a study on the bacterium *E. coli*, where 10,000 generations were studied via serial culturing (12). The resulting strains exhibited a 50% improvement in fitness as well changes in other complex phenotype such as cell size and mor-phology. Moreover, this study highlighted that the mutations regulating these phenotypes were indeed quite rare and diverse. In some cases, this change could be accomplished by point mutations; however, genomic rearrangements were also seen. Furthermore, most of the change occurred during the first 2000 generations, with improvements slowed over the last 8000. This highlights the importance of screening high levels of mutants, a prime difficulty with natural selection-based mutations. Both of these examples are highly relevant because growth improvement is a highly complex process. Not only does metabolism need to be regulated and carried out more efficiently, but a number of addi-tional factors such as substrate uptake, metabolite tolerance, and reproductive machinery also need to be optimized in a fast-growing strain. This breadth obtainable by classical strain engineering would be unfeasible in more directed approaches. As a final example, Wiebe et al. used a glucose-limited chemostat to select for mutants of *Fusarium graminearium* with delayed onset of colonial morphology, further illustrating the power of natural evolution to enrich for highly complex phenotypes (23). Delayed onset of a particular phenotype requires alteration of a wide range of regulatory factors, especially for a trait that is carried out by a plethora of cellular machinery. Furthermore, because many factors controlling morphology are unknown, directed approaches would be ineffective at generating highly improved variants.

Industrially, natural genetic drift is always under way in large-scale fer-mentations. As an example, naturally improved strains of yeasts (both

Saccharomyces and *Pichia* sp.) have been isolated from a sulfite liquor fermentation plant. These strains demonstrated the complex phenotype improvement of increased tolerance to acetic acid and enhanced galactose fermentation capacity (24). Samples from ongoing fermentations, especially long-term culturing, will present a diverse genetic population. The continuous sampling and analysis for these cultures can give rise to novel, complex phenotypes. However, other methods such as forced mutagenesis can improve the frequency of improvements as well as the prospect for success.

1.2.3 Forced Mutagenesis

Mutagenesis by ultraviolet (UV) or chemical treatment is a widely used approach for obtaining point mutations to create auxotrophic markers and improve strains. Cellular exposure to UV radiation can disrupt DNA structure, leading to a dose-specific occurrence of mutations. In addition, certain chemicals such as ethyl methane sulphonate (EMS) and nitrosomethykguanidine (NTG) have been known to cause DNA damage. It should be cautioned that all of the agents described in this section are mutagenic and hence carcinogenic and thus extreme care must be used during handling to prevent damaging exposure. In addition, certain chemicals carry orthogonal risks. For instance, NTG is explosive (25). Therefore, the benefits of increased mutation rate must be weighed against increased safety costs when working with these compounds.

The attractiveness of chemical and physical mutagens is the increased mutational capabilities compared with natural variation. Mutation frequencies are often measured as a function of auxotrophic marker development or gene mutation reversion. While basal-level, natural drift mutations can result in average reversion frequencies of nearly 1 in 10^{-8} (as described above), induced mutations by methods such as UV radiation can increase this value upwards of 10^{-3} to 10^{-5}, depending on the organism used and the intensity/duration of irradiation (26–28). Similar results and mutation frequencies can be seen with chemical mutagenesis using reagents such as EMS and NTG. Given these rates, it is still expected that the frequency of obtaining auxotrophic mutants in diploids by direct mutagenesis should be rare without prior selection. However, it has been demonstrated that auxotrophic mutants due to forced mutagenesis of diploid industrial strains can occur at frequencies of 10^{-4}, illustrating that the mechanism of mutations is still unknown (29). Therefore, improved mutants will occur at rates higher than those suggested by the probabilities. In terms of fitness landscapes, this higher mutation rate allows a further exploration of genotypic space. Thus, generating mutations with this method may yield variants located on more distant peaks, possibly at higher levels of fitness. When the phenotype of interest is influenced by a significant number of epistatic interactions, the resulting landscape will be more rugged, making forced mutagenesis more desirable for isolating improved variants than natural evolution.

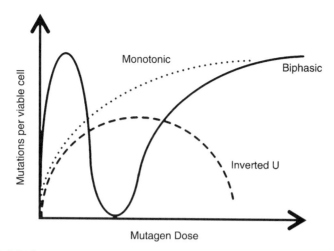

FIGURE 1.2. Common dose–response curves. Determining the optimal mutagen dose is critical for success in identifying altered mutants. This optimal level is dependent on the type/doses of mutagen and on the trait of interest. In general, three types of response curves are seen: reversion to prototrophy or resistance mutations normally follow the monotonic curve, whereas titer-increasing or decreasing mutations normally have an inverted-U shape. If the culture contains a subpopulation of radiation-sensitive individuals, biphasic behavior may be observed.

1.2.3.1 Optimal Mutagen Dose Mutation rate and cell survival are both strongly affected by mutagen dose. Thus, it is necessary to determine optimum mutagen dose. If the mutation rate is too low, variants with mutations (and especially improved phenotypes) will be rare compared with unmutated cells, making detection difficult even in high-throughput screens. In contrast, if the mutation rate is too high, the effects of deleterious mutations will swamp those of beneficial mutations, yielding poorly performing (or even nonviable) variants. Since the goal of the strain engineer is to maximize the number of beneficial mutations per variant, plots akin to Figure 1.2 are often constructed to evaluate the effects of different types or doses of mutagen on the trait of interest. A crude measure of phenotype on a small number of variants is preferred to minimize the resources spent at this preparatory stage. Curves similar to those in Figure 1.2 are often seen, depending on the phenotype of interest. Reversions to prototrophy or resistance mutations are normally monotonic, whereas titer-increasing or decreasing mutations normally follow an inverted U curve (30–32). It is important to note that as titer is improved, the likelihood of finding further beneficial mutations is reduced, making the statistic of population variance as important as the average for selection of optimal mutagen dose (33). Indeed, Lenski, et al. found a hyperbolic decline in fitness increase over time in a population of *E. coli* undergoing natural selection, suggesting that a small number of mutations of large effect were fixed in the population during early times whereas a larger number of mutations of smaller effect were fixed in subsequent generations, assuming a constant mutation rate per generation (34). Hence, the optimal mutagen dose is likely to change as superior variants are isolated. The dose that results in the highest enrichment in desirable variants is then applied and a large number of variants are screened more accurately.

Although plots akin to Figure 1.1 are useful for single-round selection programs, selecting the optimal mutagen dose based on the maximum observed phenotypic increase may cause problems for prolonged selection experiments. Although this dose will maximize the single-round phenotypic increase, subsequent rounds of improvement will have to contend with any deleterious mutations that will have occurred, possibly limiting maximal improvement in phenotype. For cases where improved titer is important, it is generally accepted that low mutation rates are better than high, even though high rates will yield a more immediate benefit (35). The rationale behind this choice is that the small number of mutations selected in a low mutation rate program will have a much higher probability of being beneficial than the high number obtained in a more error-prone program. Therefore, any subsequent beneficial mutations will not have their effects attenuated by the presence of deleterious mutations. Only after low levels of mutagen fail to yield improved variants is it desirable to increase mutagen concentration, thus expanding the evolutionary search to reveal epistatic peaks in the fitness landscape. By alternating low and high mutation rates in this manner, the prevalence of deleterious hitchhikers may be minimized.

1.2.3.2 Determination of Mutagen Specificity and Frequency A wide variety of chemical mutagens have been used to introduce DNA damage. Not only do individual chemicals produce different mutation profiles as mentioned above, but the environmental context and strain in which these mutagens are applied can also have a large effect on the changes observed (36,37). Chemical mutagens have been found to delete large (~1 kbp) sections of an organism's genome as well as generate mutations at the single base-pair level (38). Furthermore, the advent of high-throughput sequencing technology allowed the identification of mutagen-specific "hotspots" in *E. coli*, emphasizing the nonrandom nature of the induced changes (39). In addition, it has been noted that NTG acts upon the DNA replication forks, causing the resulting mutations to be tightly clustered (40). Therefore, it is highly recommended to change mutagens as a strain improvement program proceeds, not only to avoid development of resistance, but also to allow fuller sampling of genomic sequence space. Alternatively, it is possible to apply multiple mutagens in the same dose; however, the mutagens must act on different DNA repair pathways in order for this approach to be beneficial (41). In general, unless mutagenesis rates and specificities have been previously characterized for the strain of interest, characteristics of mutagens may be known only approximately, especially if the organism's cellular repair pathways are unusual.

1.2.3.3 Mechanisms of Mutagenesis Most of the mutagens introduced in this section serve to make DNA repair machinery more error prone, in addition to damaging DNA directly. Therefore, it is plausible that over the course of a selection program a mutation that confers resistance to a particular mutagen will arise. In this situation, no change in phenotype will be observed upon mutagenesis. To distinguish this case from cases where further phenotypic improvement is unlikely, some easily observable reversion phenotype may be used to confirm efficacy of the mutagenic treatment. In cases where resistance has developed it will be necessary to attempt different types of mutagens in order to introduce mutations via an alternate repair pathway (35).

UV light has been extensively studied in terms of its mutagenic frequency, specificity, and mechanism, in large part due to its ease of implementation (27,42–44). Cells may be mutated simply by exposing them to UV light for some length of time (analogous to the dose of a chemical mutagen). In the case of *E. coli*, it is thought that UV radiation causes DNA damage but that these initial lesions are not converted into base-pair changes until activation of the SOS repair pathway, a global response to DNA damage. For this reason, mutation frequency initially increases according to the square of UV dose, confirming that two distinct DNA lesions are required for mutagenesis to occur: one to induce the SOS repair pathway, and the second to cause a detectable phenotypic change. After this initial stage, mutations increase linearly with exposure as lesions continue to accumulate after SOS induction. A regime of higher order response to radiation indicates the appearance of mutations dependent on the presence of two DNA lesions in proximity (43). Finally, the mutation rate reaches a steady value as lethal mutations accumulate. Treatment with UV radiation is analogous to chemical mutagenesis in that UV has its own mutagenic specificity and frequency. However, studies have shown UV to be slightly broader in its action than other mutagens (4).

1.2.3.4 Effects of Environment Since each of the mutagens previously described require the action of cellular repair machinery, "recovery" of mutated cells in rich medium has been found to increase the mutation rate realized in the surviving cells. Not only does this treatment allow sufficient time for SOS repair to convert DNA lesions into base-pair changes, it also allows mutated proteins to be synthesized, which is important if screening occurs directly after mutagenesis. Also, certain additives to this recovery medium may promote or inhibit recovery of mutated cells. For example, addition of caffeine and acriflavine following UV mutagenesis will increase the mutation rate (35), whereas addition of manganese II, purine nucleosides, and inhibitors of protein synthesis will decrease the mutation rate (45). Interestingly, 8-methoxypsoralen is antimutagenic when present before UV irradiation, but is mutagenic if introduced after UV, illustrating the complexity of the repair mechanisms involved (46). It should be noted, however, that any treatment that increases the mutation rate will also increase lethality; therefore, it should be ascertained whether such treatments actually increase the number of mutants per survivor before mutation-amplifying additives are introduced (35).

1.2.3.5 *Applications of Forced Mutagenesis*

1.2.3.5 Applications of Forced Mutagenesis Forced mutagens have seen wide use in development of complex phenotypes. For example, UV mutagenesis was used to generate auxotrophic mutants of sake yeast (29). This is significant because it allows this yeast to be used in breeding programs and in metabolic engineering efforts utilizing molecular cloning techniques. The ability of mutagens to introduce variation in a wide variety of organisms is a major benefit to this approach, as standard genetic manipulation techniques are only established for a handful of (possibly industrially suboptimal) strains. Furthermore, EMS was used to generate sake yeast mutants with improved flavor profiles, clearly indicating the ability of forced mutagenesis to improve industrially relevant complex phenotypes whose molecular basis may be largely unknown (47,48). In addition, a forced mutagenesis/selection scheme was used to screen for improved microalgae capable of producing L-ascorbic acid. By screening over 10^5 mutants, a greater than 50-fold improvement in specific productivity was achieved (49). These results illustrate the large size of libraries necessary to achieve metabolic phenotypes. Many processes (including the penicillin production process (50)) also rely on this method to continuously enhance strains. As an example of this process improvement, penicillin titers are over 40,000-fold higher in improved strains than the original isolated wild-type strain (51). Improved antibiotic production is a prime example of a complex phenotype, as strains must evolve not only the enzymes responsible for antibiotic synthesis but also any factors involved in nutrient transport and chemical tolerance. The organisms normally responsible for high production of a compound of interest are often genetically uncharacterized; thus, the genome-blind nature of the classical approach becomes an asset. This approach of mutagenesis and screening has even been used to improve the activity of baker's yeast for bread making purposes (52). As a result of the ease of operation and selection, this method continues to be used to generate complex phenotypes in industrial cells.

1.2.4 Strain Mating

Strain mating represents an effective tool for generating a population with a high number of non-detrimental mutations. One of the main limitations of random mutagenesis is the high probability that the changes induced in a daughter cell will be detrimental, and this probability increases as the mutation rate increases. However, high mutation rates are required to escape local optima on the fitness landscape. Strain mating allows recombination to occur between two divergent (yet functional) genotypes, generating a library of highly mutated individuals. However, because meiotic recombination operates at the level of the gene, as opposed to the base pair, any mutations introduced are likely to be in the genomic context in which they were found in one of the parents, decreasing the likelihood that such mutations are lethal. In essence, this technique allows exploration of distant regions of the fitness landscape without the associated high probability of failure, thus allowing generation of mutants located specifically on regions of high fitness. It is obvious that strain mating will have its maximum effect when applied to two highly divergent members of the same species, allowing a high rate of mutation with a low probability of lethality. This technique allows the possibility of running multiple mutagenesis programs in parallel (utilizing perhaps different mutagenic techniques and screening strategies) and mating the most successful individuals from each program, especially if mutagenesis has been conducted to minimize the occurrence of deleterious mutations. In fact, it has been theorized that the accumulation of deleterious mutations causes evolving populations to gradually reach a maximum fitness. If this is the case, then strain mating should greatly improve the potential for phenotypic increase (53). It should be cautioned, however, that if significant epistatic interactions exist between genes, there will be a high likelihood of disrupting them upon mating, possibly leading to inferior individuals (54).

Protoplast fusion is a distinct method of strain mating that does not involve recombination. Instead, the cell walls of two individuals are digested away and their genetic material is combined to form a new individual with both sets of chromosomes (55). This technique allows the characteristics of both organisms to be combined (forming a heterokaryon) without the risk of recombination loss. Heterokaryons are often verified by nutrient complementation. Therefore, it is imperative that each parent be auxotrophic for a different compound. Removing the cell wall while preserving the cell membrane is a delicate process. Therefore, reliable isolation of heterokaryons is dependent on a number of factors, including protoplast isolation from exponentially growing cells, maintenance of isotonicity in the protoplast media, and the addition of polyethylene glycol as a fusogenic agent. Since both genomes are isolated from viable individuals, the probability of deleterious interactions is small. Further, since enzymatic deficiencies are recessive, any lack of functionality in one parent will be complemented by the genome of the other. An added benefit of generating a polyploid strain is the doubling of the effective mutation rate for each gene, allowing evolution of improved phenotypes to proceed at a faster pace, as mentioned earlier.

Strain mating can be used to combine two distinct functionalities into one organism. For example, a common problem in fermentations is the evolution of a "killer" phenotype, whereby a nonproductive individual gains the ability to secrete a toxic compound, thus outcompeting the organisms of desirable phenotype and resulting in a failure of the fermentation. However, strain mating can be used to generate a productive "killer" phenotype, whereby the resulting population is able to both secrete the product of interest and kill any contaminants that may be introduced (56). Indeed, Bortol et al. were able to fuse strains of *S. cerevisiae* possessing the "killer" phenotype with traditional baker's yeast, producing competitive variants that retained the ability to make dough rise (57). Clearly, strain mating has enormous potential for generating mutants improved in a variety of complex phenotypes.

1.3   GENOTYPIC LANDSCAPES

If phenotype and genotype are graphed such that related genotypes are close together, the resulting landscape is ripe with series of peaks and valleys, with peaks representing genotypes of high fitness and valleys representing geno-types of low fitness. Natural selection dictates that individuals residing on higher peaks are more likely to reproduce, and upon reproduction, a new generation arises at some genotypic distance away from their parents, depend-ing on the mutation rate. As evolution proceeds, the population will tend toward peaks and away from valleys if the selection pressure is toward improved phenotype. As a result, understanding aspects of this landscape helps develop proper mutagenesis and selection strategies. For example, what magnitude of fitness differential is observed when moving from a peak to a valley? Are there many different peaks, or just one? Do there exist mountain ridges connecting each peak, or are each separated by deep chasms? A good understanding of the topography of this fitness landscape will allow prediction of the evolutionary trajectories of a population under selection. For the strain engineer, this understanding will allow comparison of different selection pro-grams in terms of their ability to generate a mutant residing on the tallest peak of this evolutionary landscape.

One important quality of fitness landscapes is their "ruggedness" as devel-oped by Kauffman (58). This quality indicates the correlation in phenotype observed between related genotypes. In the limit of no ruggedness (i.e., perfect correlation), it can be shown that there is only one peak in the fitness land-scape and that this peak is accessible from any genotype by progressing through successively more fit one-mutant neighbors. However, in the limit of maximal ruggedness (no correlation between related genotypes), the land-scape is essentially random, with many local optima and a very small chance of encountering the global optimum by progressing through successively more fit one-mutant neighbors. This quality is therefore extremely important for evaluating *a priori* which mutagenic and selection treatments are likely to yield improved mutants. Treatments that result in small genotypic changes (such as single base-pair changes) may only be able to proceed in small steps through the fitness landscape. If selection is operated such that the fittest mutant is selected for subsequent mutagenesis and selection, then repeated rounds of generation and screening are only able to yield a local maxima. However, as mutagenic treatments become more severe, the possibility of generating a mutant on a more distant peak becomes higher at the expense of mutant generation on the current peak. Naturally, more severe mutagenic treatments become more desirable as the number of peaks in the evolutionary landscape increases. Alternatively, selection regimes that preserve a nonzero fraction of suboptimal mutants may also have an advantage in detecting more distant peaks. However, the costs associated with running many selection programs in parallel may prove too great (59). It has been shown that phenotypes that exhibit a high degree of epistasicity resemble more rugged fitness landscapes, whereas phenotypes under the control of genotypes whose effects are perfectly additive resemble the gradual "Fujiyama" type of landscape (58).

1.4 SCREENING

The success of classical strain engineering is due in large part to the ability of researchers to search through a large number of variants to isolate a few improved individuals: a process called screening. As more mutants are screened, the probability of isolating an improved variant increases linearly (assuming the number of mutants generated is large compared with the number of mutants screened). Therefore, significant effort has been spent to develop improved techniques and technologies to allow larger numbers of variants to be assayed per unit time. Screens can be classified into one of two broad categories: rational screens and random screens (35). Rational screens are defined by their exploitation of knowledge about the system of interest, whereas random screens are of more general applicability. Ideally, the quantity being measured during the screen will correspond exactly to the phenotype of interest, but in cases where this is difficult to measure several orthogonal correlates of phenotype may be assayed to decrease the rate of false positives. It is important to keep in mind that the optimal screening strategy will depend in large part on the phenotype of interest and any prior knowledge of the system. In addition, resource limitations may restrict which screens may be performed as well as the number of mutants that may be assayed. The error rate of a screen is also of critical importance, as poorly designed growth screens may yield false positives and noisy assays will necessitate screening replicates to increase confidence. Finally, the importance of selecting individuals that exhibit true phenotypic improvements cannot be overemphasized. The powerful techniques of classical strain engineering often generate individuals that may perform well in a particular screen yet do not produce the phenotype of interest. Although detection of these "screening artifacts" can be largely eliminated through careful experimental design, further characterization of isolates at the conditions of interest is often required.

1.4.1 Rational Screens

In general, there are at least as many ways to screen for a particular phenotype as there are measurable phenotypes. However, a handful of rational screening strategies stand out due to their popularity and generality. It should be noted that all rational screening procedures assume at least a crude knowledge of the mechanism by which a phenotype is manifested. For instances in which this is not well known, it may be necessary to proceed first with a random screen to identify improved variants, followed by a study to determine which screens are most selective for the isolated individuals. However, it should be noted that none of the techniques mentioned below assume a *molecular* knowledge of the biochemistry involved, which is the minimum requirement for a directed approach to succeed, indicating this approach's generality for a wide variety of uncharacterized microbial strains and complex phenotypes.

Phenotypic titer depression is a common way of shifting the "detectable range" of a random or rational screen. It is often the case that one is interested in mutants exhibiting a high rate of product secretion or growth rate. However, the method used to detect phenotypic changes may not be accurate at the range of interest, especially when the population under selection is derived from a highly improved parent. Therefore, by artificially decreasing titer, differences among high-producing variants may be discerned. This is often accomplished by altering media composition so that a particular nutrient is limiting product formation, or through introduction of a metabolic inhibitor. It is assumed that individuals proficient under these limiting conditions will maintain their superiority in a production setting (60).

Toxic analogs of metabolic precursors can be used to select for variants with improved metabolic qualities. When a metabolic precursor is synthesized intracellularly, mutants resistant to its toxic analog may be overproducing the nontoxic compound, diluting the poisonous effects of the analog and increasing flux through the pathway of interest. This method has been applied successfully for bioproducts derived from amino acids (61). For compounds provided as nutrients in the growth medium, however, sensitivity to their toxic analog may indicate improved transport properties for that class of molecule, thus increasing metabolic flux toward the pathway of interest. One potential drawback of this method (and assaying for sensitivity in general) is that it must be accomplished through replica plating, which has much lower throughput and is more labor-intensive than screening in liquid culture (62).

In instances where the product of interest is known to inhibit the activity of a toxic compound, selection for resistant mutants may result in isolates of improved production (62). This screening method, called selective detoxification, is most applicable to solid media due to its ability to provide each mutant with a unique chemical environment; liquid cultures allow the product of interest to diffuse and provide resistance to nonproducers, confounding results. It should be cautioned, however, that if alternate pathways to resistance are present, the possibility of encountering screening artifacts may be unavoidable. This method has seen success in generating *Acremonium chrysogenum* variants proficient in detoxifying metallic ions through production of Cephalosporin C (61).

Desirable concentrations of the product of interest may be infeasible for a number of reasons. First, the desired compound may be directly toxic to the cell. Second, the product of interest may participate in an inhibitory feedback loop, which limits its production. The first bottleneck may be alleviated simply by screening for individuals resistant to high concentrations of the desired compound. Mutants deficient in feedback inhibition may be isolated by screening for mutants resistant to a toxic analog of the end product. It is expected that survivors will be deregulated, overproducing the compound of interest and thus diluting the effect of the toxic analog (63).

A particularly clever screening strategy involves the mutagenesis of non-producing strains, isolated through mutagenesis of a productive parent strain. In theory, productive mutants isolated after this second round of mutagenesis will have had at least two mutations in the relevant biosynthetic genes: an inactivating mutation followed by a mutation that restores productive ability to levels that are (hopefully) higher than the parent strain. An added benefit of this method is the low level of background activity observed, enabling more rapid screening techniques to be employed. Furthermore, revertants are more likely to contain mutations in genes directly related to product synthesis, as opposed to genes whose effect is epistatic (62). This technique has seen success in overproduction of the antibiotic aurodox in *Streptomyces goldiniensis* (63).

As a strain of interest becomes more highly optimized, the likelihood of generating phenotypic changes of large magnitude steadily decreases. Hence, the maximum expected improvement in phenotype may be within the error of the screen. To increase the probability of detecting variants with low (but significant) improvement, a rapid recycling scheme can be implemented (35,59). In this statistical approach, a large rake-off (~10–50%) of mutants are immediately rescreened. This process is repeated multiple times to enrich the fraction of genuinely improved variants, the rate of enrichment corresponding to the magnitude of phenotypic increase. Mutagenesis can be undertaken between rounds of recycling or after isolation and characterization of improved individuals. Due to the power of this technique, screening artifacts can become a major concern if the selective conditions are poorly designed. Although such statistical rigor is recommended throughout the optimization process, it becomes critical to continued isolation of improved variants as phenotypic increases become more marginal and rarer.

A significant number of phenotypes cannot be linked to microbial growth, necessitating the development of alternate screening methodologies. Colorful or fluorescent phenotypes may be detected spectrometrically (9,64), but for phenotypes that do not exhibit obvious color or fluorescence, a substantial amount of creativity is often required. Identification of a suitable colorimetric assay may be relatively simple for popular phenotypes, but in more specialized cases a solution may have to be developed in-house. In any instance where a large amount of processing is necessary before a phenotype can be measured, screen throughput will be significantly diminished and optimization of assay protocols becomes of paramount importance.

1.4.2 Random Screens

In the absence of any knowledge about the causative factors of the phenotype of interest, a random screen is often the only option for isolating desirable variants. However, the conditions of the screen must be very similar to those of the final production setting of these strains or else screening artifacts will be encountered. The major concern with random screening is the immense library size and screening effort required.

A common way of quickly reducing this library size and isolating interesting variants is known as preselection. In this approach, a crude growth-based correlate of the phenotype of interest is used to eliminate any variants that are not superior to the parent strain. This scheme is especially useful in cases where accurate measures of phenotype are difficult to achieve, thus precluding their use in the entire mutated population. In cases where the phenotype of interest naturally confers a growth advantage, preselection can simply consist of a crude growth-based random screen. Otherwise, when interested in the production of a secondary metabolite, any of the rational screens discussed above may be used (61). Since the aim of a prescreen is to increase the throughput of a selection program, the time savings conferred by the prescreen must be sufficient to make its inclusion worthwhile (59).

1.4.3 Screening Platforms

When the phenotype of interest can be directly coupled to growth, selection based on growth rate offers a simple, high-throughput method for isolating improved variants. Growth conditions are of critical importance in such schemes, as poor choices will result in a high incidence of screening artifacts. In addition to the chemical environment in which selection takes place, the physical environment will also have a significant impact upon which mutants exhibit a growth advantage. The physical environments most commonly used include agar plates, batch culture, and continuous culture.

1.4.3.1 Solid Media The defining feature of solid media for microbial growth is its resistance to diffusion. Not only are individual variants spatially separated, but also any diffusible metabolites remain localized to their parent colony. As mutants are spatially separated, they do not compete with one another for nutrients, allowing individuals to be isolated, maximizing the phenotype of interest, as opposed to those who use energy to decrease the fitness of other mutants. Additionally, mutants exhibiting significant growth differences are easily discernible by eye or by image processing software. Also, differential secretion of a colorful or bactericidal compound can be identified by the size of "halos" surrounding each colony. However, since colony diameter increases as the cube root of population and halo diameter as the square root of secretion capability, differences among high-producing individuals may not be discernible. To overcome this limitation, phenotypic titer depression, as discussed earlier, may be implemented (60). However, phenotypic advantages in these artificial conditions may not translate to an advantage in a production setting. Indeed, growth conditions on agar plates in general are significantly different from those present in a bioreactor, and as such, testing under more realistic conditions is often necessary to refine the pool of promising individuals. An additional consideration when screening on solid media is the maximum allowable throughput. Although 1 mL of liquid media may contain upwards of 10^8 individuals, a 100-mm plate may only contain 10^3–10^4 in order to allow sufficient time for the phenotype to be expressed before colonies become indistinguishable. Furthermore, high plating densities on selective media may decrease the recovered fraction of mutants due to the Grigg effect (65,66). Briefly, plating a high density of nonviable cells may inhibit the growth of viable ones due to nutrient consumption or secretion of a toxic compound. Therefore, the benefit of colony separation must be weighed against increased throughput when designing such a growth-based screen. Screening programs incorporating agar plates have been used effectively to select for a variety of highly complex phenotypes, including antibiotic production (67), amino acid auxotrophy (68), ethanol production (69), as well as numerous improved tolerance applications.

1.4.3.2 Batch Culture Batch culture is characterized by repeated cycles of exponential and stationary phase growth. Therefore, variants under selection are alternately subjected to rich and starvation conditions. Those mutants that can reproduce the fastest under rich conditions will be preferentially selected as colonists of the next batch culture. Given the exponential nature of bacterial growth, mutants with even a slight growth advantage will come to dominate the final population. Hence, this environment is best suited for isolating strains with reduced lag time and higher growth rates (70). However, growth is essentially the only phenotype that may be selected for using this approach. Since variants are not spatially separated and secreted compounds are freely diffusible, mutants cannot be distinguished based upon their secretory characteristics. Additionally, any mutant that secretes a toxic compound to which the mutant itself is immune will have a selective advantage unrelated to the screen's intended phenotype. These "killer" phenotypes, although seemingly inconvenient, offer the ability to confer a selective advantage to a production strain (through a technique such as protoplast fusion), extending the time over which a fermentation may take place before contamination occurs (56). It cannot be overemphasized that the selective environment encountered in liquid culture is highly dependent on the microbial ecology. Unlike in solid media, individuals in liquid culture continuously compete for the same nutrients. Hence, selective conditions will change with time as microbial populations change and the superiority of selected variants will, in general, be dependent upon the microbial environment in which they were grown. In other words, the fittest variant among a competitive population may not be superior when considered in isolation. On the other hand, liquid media provides an excellent environment for optimization programs using an organism's natural mutation rate, as improved variants are continuously being generated and taking over the existing mutant pool. Perhaps the best known example of such a long-term evolution experiment comes from Lenski et al., who subjected *E. coli* to batch conditions for 10,000 generations. It was found that individuals present at the end of the experiment had a shorter lag phase and higher growth rate than the strain used to start the experiment. In addition, it was found that most of the competitive advantage was obtained within the first 2000 generations of the culture (34). Although Lenski et al. were not interested in generating an industrially useful phenotype, these results imply that similar techniques would be very effective at generating improved isolates of industrial relevance. Indeed, through cycled batch cultivations of *S. cerevisiae* in glucose, xylose, and arabinose, a variant that obtained the ability to completely ferment all three sugars in almost half the time as the parent was isolated (71). Such an improvement would require a highly detailed understanding of the bottlenecks limiting the consumption of each sugar, including transport, metabolism regulation (to alter diauxie) and carbon metabolism. Characterization of each of these components (not to mention analyzing their interaction) would be an enormous undertaking if a directed approach were to be followed. However, by simply allowing faster-growing mutants to outcompete less fit individuals, a highly desirable solution to this complex problem can be achieved.

1.4.3.3 Continuous Culture

Chemostats, in contrast to batch cultures, operate at steady state, with a steady outflow of culture balanced by a corresponding influx of media (at a level below that which would wash out all of the cells). Those individuals that are best able to utilize these low levels of nutrients will have a selective advantage under this condition. Hence, instead of selecting for mutants with a high μ_{max}, as is the case in batch cultures, chemostats select for variants with a low K_s, that is, the concentration of a limiting nutrient (such as glucose) at which the growth rate of an organism achieves half its maximal value. Thus, chemostats tend to select for specialists who can make maximum use of a limiting nutrient instead of selecting for general opportunists of high growth rate, as for batch cultures. In addition, because all individuals share the same nutrient pool, the possibility of forming stable ecologies exists, with the unused nutrients and excreted metabolites of one population providing nutrients for a second (72). This situation, though problematic for instances requiring monoclonal cultures (e.g., when protoplast fusion with another variant is desired), may be acceptable in other cases (e.g., remediation of a toxic compound). When the chemostat population is largely monoclonal, however, evolution in a chemostat follows a strictly sequential process, with fitter variants deriving from the most populous clone and subsequently replacing it. Since fitness differences are not transitive (due to epistatic effects), it is possible for the fitness of a population to decrease with time, as measured by pairwise comparisons between isolates that are not immediately related to one another (73). In these cases, there is no "best" variant for a particular selective environment due to fitness's dependence on the microbial composition of the chemostat culture. In addition, since chemostats select for populations with high residence time, adherence to bioreactor walls can become a major concern (70). Finally, in comparison to agar plates and batch cultures, the chemostat apparatus can be quite expensive. Nevertheless, chemostats have been quite successful in the development of a wide variety of very complex phenotypes, with results such as altered morphology (23), increased plasmid stability (74), and increased xylose uptake rate (22).

Although the steady-state operation of the chemostat is desirable for some selection programs, it is often necessary to control a particular variable as growth proceeds. The combination of a chemostat with an online controller is referred to as an auxostat. The increased versatility of this instrument (reflected in its increased cost) allows a wide variety of schemes to be implemented during screening. For example, by controlling media flow rate to maintain a constant low cell density, a strong pressure can be applied to select mutants with high growth rate in rich media. In essence, this setup results in a batch reactor with infinite volume, which is useful when the ability to adapt to stationary phase is not required (75). Alternatively, instead of varying the media flow rate to control cell density, the concentration of a toxic compound may be steadily increased, resulting in selection for a tolerant phenotype. The process of continuously changing selective conditions in real time as improved variants emerge is known as interactive continuous selection (ICS) (76). This method has been employed to select for *Streptomyces* mutants tolerant to increasingly high levels of streptomycin, resulting in strains that produce large quantities of this antibiotic (76). Finally, it is possible to simulate a continuous culture through serial batch subculturing in which a fraction of a batch culture is reinoculated into a fresh culture. The growth state of the inoculum and frequency of transfer will dictate how similar the process will be to either batch or continuous.

1.4.3.4 Modern Screening Platforms   Although liquid culture allows for a much higher numbers of variants to be screened than solid media, it allows crosstalk between individuals, altering the selective pressure applied to the cells. To overcome this limitation, Naki et al. developed a microtube-based screening system that allows a growth-based selection to be applied in liquid

media while preventing crosstalk (77). It is estimated that this method can provide an order of magnitude increase in throughput as compared with solid media.

When the phenotype of interest results in a visible change at the single-cell level, microfluidic techniques allow multiple orders of magnitude improvement in throughput over other non-growth-based assay systems. In particular, 10^8 mutants per hour may be assayed via flow cytometry, which queries individual cell size and fluorescence. One issue unique to such a sensitive instrument is the ability to detect significant variability among a monoclonal population. Hence, it is possible to isolate what seems like improved variants that reproduce to yield an unimproved population average. Thus, characterization of average cell-to-cell variability is of paramount importance in designing a precise flow cytometric screen. Nevertheless, mutations that increase cell variability may arise, necessitating the use of a rapid recycling scheme allowing cell division between each measurement. Of course, the phenotype of interest must result in a visible difference at the single-cell level, but nevertheless flow cytometry has been successfully employed to enrich for a wide variety of phenotypes. For example, carotenoids exhibit a characteristic fluorescence and are localized to cellular membranes, thus allowing An et al. to select for yeast variants with improved carotenoid production capabilities (9). Furthermore, Tyo et al. implemented a product-specific stain to select for overproducers of poly-3-hydroxybutyrate, a thermoplastic of commercial relevance (64).

Despite efforts to adapt the selection of many phenotypes to high-throughput platforms such as growth cultures or solid media, it is often necessary to screen variants individually in liquid media. For example, secretion of a particular compound may not occur on solid media, or it might be desirable to test isolates obtained via another method under more industrially relevant conditions. In addition, a large cell count may be necessary for more accurate phenotype quantification. In instances where development of a phenotype in 50-mL shake flasks would be too resource-intensive, deep 96-well plates offer a reasonable compromise. Isolates may be grown in up to 2 mL of media in plates especially designed to maximize aeration and prevent cross-contamination (78). Depending on the phenotype of interest, up to 10^4 variants may be assayed per day per technician. Finally, the 96-well format has gained wide acceptance in industry, prompting the development of a plethora of equipment specifically designed for running experiments in this setting.

The development of robotics and microcontrollers during the past 50 years has greatly enhanced the efficiency of selection programs, especially for cases when variants must be kept separate. Screens based on solid media can greatly benefit from automated colony pickers equipped with image analysis software. In addition, more specialized systems exist for inoculating a lawn of bacteria with "plugs" from a plate containing antibiotic-secreting variants to determine inhibition zones and hence product secretion ability. Furthermore, a wide variety of robotic systems designed for manipulation of cultures in the 96-well plate format have been developed, including media handlers, plate movers, plate storage systems, and plate readers. One important consideration when operating a robotic system is maintenance of sterility, as robotic components comes into regular contact with a large number of cultures. Additionally, robotic screening systems are only as good as their software; interesting or unexpected phenotypes will not be selected unless their characteristics have been programmed into the detection routine. Despite the added complexities associated with operating a robotic system, expenditure of a reasonable amount of care will make the operation of a high-throughput, statistically rigorous screening program much more efficient. (For more information on high throughput fermentation techniques, see Chapter 5.)

1.5 CONCLUSIONS

Complex phenotype optimization via the classical approach is well established in the food and pharmaceutical industry. Improvement of yeast strains for alcohol fermentations has long taken the classical approach due to "generally recognized as safe" (GRAS) classification and ease of selection. This approach has been quite successful in improving complex phenotypes such as complex metabolite profiles, flocculation, and chemical tolerances (79). The success of this approach can be seen in the evolution of the sake fermentation yeast (See Box 1.1). In addition, since the advent and discovery of antibiotics, a long-standing goal has been the increase of titer. The significant improvements seen in these processes have mostly been due to the use of the classical strain engineering approach (see case study in Chapter 6).

The genome-wide mutations induced by classical strain engineering are not as efficient when the desired mutations occur in a single gene. However, when it is desirable to obtain mutations across many genes in the cell (as is often the case with complex phenotypes), the global nature of this approach is an asset. Moreover, there is no need to understand the underlying genetic and regulatory network to direct mutagenesis, as the "space" of possible mutations covers the entire genome, in contrast to rational methods, which require more intimate knowledge of influential genes to be successful. Classical strain engineering, therefore, may return mutants that exploit previously unknown regulatory mechanisms or metabolic pathways, making this approach applicable not only to organisms that are poorly characterized, but also to model organisms. Furthermore, techniques of classical strain engineering can induce previously dormant sections of a genome to become active. Exploitation of these "cryptic genes" would be unlikely in a rational approach to strain improvement, demonstrating the ability of classical strain engineering to find novel and nonintuitive solutions to a design goal (80). An important disadvantage of this method is that the incurred changes are not easily traceable or movable to another host strain. Recently, advances in whole-genome resequencing and "omics" technologies are beginning to evaluate these strains in hopes of identifying the underlying changes (11) (see Chapter 3). However, this sort of inverse metabolic engineering is seen to be a new frontier at the interface of the classical and rational approaches for complex phenotype engineering (81).

The classical strain engineering approach has long stood the test of time in the fermentation industry due to its ability to consistently generate improved phenotypes using simple techniques. By starting from single base-pair changes and progressively increasing the rate of mutation, the strain engineer can explore ever-more distant reaches of the fitness landscape, eventually traversing wide valleys in single bounds as optimized strains are combined to create individuals for further mutagenesis and improvement. The power of these techniques to improve complex phenotypes lies in the lack of assumptions made in their application. No hypotheses about rate-limiting steps or flux imbalances are needed to generate improved variants, just a well-designed assay and patience. Luckily, with the continued introduction of cost-effective robotic and microfluidic systems, the length of time required for isolation of improved variants will steadily decrease. Further, the use of this technique is readily accepted by both regulators and consumers for the improvement of food organisms. The generality of this approach, however, is often its major downfall. Rational metabolic engineering, with its ability to precisely alter the function of specific genes, is often able to generate improved variants in much less time than classical strain engineering when such detailed knowledge is available (see Chapter 2). Furthermore, the directed nature of such rational techniques allows inferences to be made about the mechanism underlying a phenotype, even when such techniques do not work. On the contrary, successful variants isolated through classical techniques cannot yield any information about underlying causes. As genome sequencing continues to increase in speed and affordability, however, the ability to uncover and rationalize the causes of phenotypic increase in classically engineered variants will increase. Thus, classical techniques promise not only to continue to yield improved strains, but also to elucidate the hidden bases of complex phenotype display in microorganisms.

REFERENCES

1. Doebley, J. (2004) The genetics of maize evolution. *Annu Rev Genet*, **38**, 37–59.
2. Demain, A. and Davies, J. (1998) *Manual of Industrial Microbiology and Biotechnology*, 2nd ed. ASM Press, Washington, DC.
3. Parekh, S., Vinci, V.A., and Strobel, R.J. (2000) Improvement of microbial strains and fermentation processes. *Appl Microbiol Biotechnol*, **54**, 287–301.
4. Cupples, C.G. and Miller, J.H. (1989) A set of lacZ mutations in *Escherichia coli* that allow rapid detection of each of the 6 base substitutions. *Proc Natl Acad Sci U S A*, **86**, 5345–5349.

5. Hampsey, M. (1991) A tester system for detecting each of the 6 base pair substitutions in *Saccharomyces cerevisiae* by selecting for an essential cysteine in iso-1-cytochrome C. *Genetics*, **128**, 59–67.

6. Cupples, C.G., Cabrera, M., Cruz, C., and Miller, J.H. (1990) A set of lacZ mutations in *Escherichia coli* that allow rapid detection of specific frameshift mutations. *Genetics*, **125**, 275–280.

7. Patrick, W.M., Firth, A.E., and Blackburn, J.M. (2003) User-friendly algorithms for estimating completeness and diversity in randomized protein-encoding libraries. *Protein Eng*, **16**, 451–457.

8. Firth, A.E. and Patrick, W.M. (2008) GLUE-IT and PEDEL-AA: new programmes for analyzing protein diversity in randomized libraries. *Nucleic Acids Res*, **36**, W281–W285.

9. An, G., Bielich, J., Auerbach, R., and Johnson, E. (1991) Isolation and characterization of carotenoid hyperproducing mutants of yeast by flow cytometry and cell sorting. *Biotechnology*, **9**, 70–73.

10. Fan, Z., McBride, J.E., Zyl, W.H.V., and Lynd, L.R. (2005) Theoretical analysis of selection-based strain improvement for microorganisms with growth dependent upon extracytoplasmic enzymes. *Biotechnol Bioeng*, **92**, 35–44.

11. Minty, J., Lesnefsky, A., Lin, F., Chen, Y., Zaroff, T., Veloso, A., Xie, B., McConnell, C., Ward, R., Schwartz, D. et al. (2011) Evolution combined with genomic study elucidates genetic bases of isobutanol tolerance in *Escherichia coli*. *Microb Cell Fact*, **10**, 18.

12. Papadopoulos, D., Schneider, D., Meier-Eiss, J., Arber, W., Lenski, R.E., and Blot, M. (1999) Genomic evolution during a 10,000-generation experiment with bacteria. *Proc Natl Acad Sci U S A*, **96**, 3807–3812.

13. Berg, O.G. (1995) Periodic selection and hitchhiking in a bacterial population. *J Theor Biol*, **173**, 307–320.

14. Dykhuizen, D.E. (1990) Experimental studies of natural selection in bacteria. *Annu Rev Ecol Syst*, **21**, 373–398.

15. Paquin, C. and Adams, J. (1983) Frequency of fixation of adaptive mutations is higher in evolving diploid than haploid yeast populations. *Nature*, **302**, 495–500.

16. Degnen, G.E. and Cox, E.C. (1974) Conditional mutator gene in *Escherichia coli*: isolation, mapping, and effector studies. *J Bacteriol*, **117**, 477–487.

17. Nghiem, Y., Cabrera, M., Cupples, C.G., and Miller, J.H. (1988) The mutY gene: a mutator locus in *Escherichia coli* that generates GC->TA transversions. *Proc Natl Acad Sci U S A*, **85**, 2709–2713.

18. Taddei, F., Radman, M., MaynardSmith, J., Toupance, B., Gouyon, P.H., and Godelle, B. (1997) Role of mutator alleles in adaptive evolution. *Nature*, **387**, 700–702.

19. Dykhuizen, D.E. (1993) Chemostats used for studying natural selection and adaptive evolution. *Methods Enzymol*, **224**, 613–631.

20. Herbert, D., Elsworth, R., and Telling, R.C. (1956) The continuous culture of bacteria; a theoretical and experimental study. *J Gen Microbiol*, **14**, 601–622.

21. Jansen, M.L.A., Diderich, J.A., Mashego, M., Hassane, A., de Winde, J.H., Daran-Lapujade, P., and Pronk, J.T. (2005) Prolonged selection in aerobic, glucose-limited

chemostat cultures of *Saccharomyces cerevisiae* causes a partial loss of glycolytic capacity. *Microbiology*, **151**, 1657–1669.

22. Kuyper, M., Toirkens, M.J., Diderich, J.A., Winkler, A.A., van Dijken, J.P., and Pronk, J.T. (2005) Evolutionary engineering of mixed-sugar utilization by a xylose-fermenting *Saccharomyces cerevisiae* strain. *FEMS Yeast Res*, **5**, 925–934.

23. Wiebe, M.G., Robson, G.D., Oliver, S.G., and Trinci, A.P.J. (1994) Use of a series of chemostat cultures to isolate improved variants of the Quorn(R) mycoprotein fungus, Fusarium graminearum A3/5. *Microbiology (UK)*, **140**, 3015–3021.

24. Linden, T., Peetre, J., and Hahn-Hagerdal, B. (1992) Isolation and characterization of acetic acid-tolerant galactose-fermenting strains of *Saccharomyces cerevisiae* from a spent sulfite liquor fermentation plant. *Appl Environ Microbiol*, **58**, 1661–1669.

25. Bridges, B. (1976) In Macdonald, K. (ed.), *Genetics of Industrial Microorganisms 1974*. Academic Press, London, Vol. 2, pp. 7–14.

26. Doudney, C.O. and Rinaldi, C.N. (1984) Modification of UV-induced mutation frequency and cell survival of *Escherichia coli* B/r WP2 trpE65 by treatment before irradiation. *J Bacteriol*, **160**, 233–238.

27. James, A.P. and Kilbey, B.J. (1977) The timing of UV mutagenesis in yeast: a pedigree analysis of induced recessive mutation. *Genetics*, **87**, 237–248.

28. Rajpal, D.K., Wu, X., and Wang, Z. (2000) Alteration of ultraviolet-induced mutagenesis in yeast through molecular modulation of the REV3 and REV7 gene expression. *Mutat Res*, **461**, 133–143.

29. Hashimoto, S., Ogura, M., Aritomi, K., Hoshida, H., Nishizawa, Y., and Akada, R. (2005) Isolation of auxotrophic mutants of diploid industrial yeast strains after UV mutagenesis. *Appl Environ Microbiol*, **71**, 312–319.

30. Chang, L.T., Lennox, J.E., and Tuveson, R.W. (1968) Induced mutation in UV-sensitive mutants of *Aspergillus nidulans* and *Neurospora crassa*. *Mutat Res*, **5**, 217–224.

31. Brown, W.F. and Elander, R.P. (1966) Some biometric considerations in an applied antibiotic AD-464 strain development program. *Dev Ind Microbiol*, **7**, 114–123.

32. Mehta, R. and Weijer, J. (1971) *Radiation and Radioisotopes for Industrial Microorganisms*. International Atomic Energy Agency, Vienna.

33. Orr, H.A. (1999) The evolutionary genetics of adaptation: a simulation study. *Genet Res*, **74**, 207–214.

34. Lenski, R.E., Mongold, J.A., Sniegowski, P.D., Travisano, M., Vasi, F., Gerrish, P.J., and Schmidt, T.M. (1998) Evolution of competitive fitness in experimental populations of *E. coli*: what makes one genotype a better competitor than another? *Antonie Van Leeuwenhoek*, **73**, 35–47.

35. Rowlands, R. (1983) In Smith, J., Berry, D., and Kristiansen, B. (eds.), *The Filamentous Fungi*. Edward Arnold, London, Vol. 4, pp. 346–372.

36. Prakash, L. and Sherman, F. (1973) Mutagenic specificity: reversion of iso-1-cytochrome C mutants of yeast. *J Mol Biol*, **79**, 65–82.

37. Coulondre, C. and Miller, J.H. (1977) Genetic studies of lac repressor: IV: mutagenic specificity in lacI gene of *Escherichia coli*. *J Mol Biol*, **117**, 577–606.

38. Liu, L.X., Spoerke, J.M., Mulligan, E.L., Chen, J., Reardon, B., Westlund, B., Sun, L., Abel, K., Armstrong, B., Hardiman, G. et al. (1999) High-throughput isolation of Caenorhabditis elegans deletion mutants. *Genome Res*, **9**, 859–867.

39. Parkhomchuk, D., Amstislavskiy, V., Soldatov, A., and Ogryzko, V. (2009) Use of high throughput sequencing to observe genome dynamics at a single cell level. *Proc Natl Acad Sci U S A*, **106**, 20830–20835.

40. Cerda-Olmedo, E. and Ruiz-Vazquez, R. (1979) In Sebek, O. and Laskin, A. (eds.), *Genetics of Industrial Microorganisms 1978*. American Society for Microbiology, Washington, Vol. 3, pp. 15–20.

41. Talmud, P. (1977) Mutational synergism between para-fluorophenylalanine and UV and Coprinus lagopus. *Mutat Res*, **43**, 213–222.

42. Cohen-Fix, O. and Livneh, Z. (1992) Biochemical analysis of UV mutagenesis in *Escherichia coli* by using a cell-free reaction coupled to a bioassay: identification of a DNA repair-dependent, replication-independent pathway. *Proc Natl Acad Sci U S A*, **89**, 3300–3304.

43. Doudney, C.O. (1976) Complexity of ultraviolet mutation frequency-response curve in *Escherichia coli* B/r: SOS induction, one-lesion and two-lesion mutagenesis. *J Bacteriol*, **128**, 815–826.

44. Rajpal, D.K., Wu, X.H., and Wang, Z.G. (2000) Alteration of ultraviolet-induced mutagenesis in yeast through molecular modulation of the REV3 and REV7 gene expression. *Mutat Res-DNA Repair*, **461**, 133–143.

45. Clarke, C.H. and Shankel, D.M. (1975) Antimutagenesis in microbial systems. *Bacteriol Rev*, **39**, 33–53.

46. Bridges, B.A. (1971) Genetic damage induced by 254 nm ultraviolet light in *Escherichia coli*: 8-methoxysporalen as protective agent and repair inhibitor. *Photochem Photobiol*, **14**, 659–662.

47. Ichikawa, E., Hosokawa, N., Hata, Y., Abe, Y., Suginami, K., and Imayasu, S. (1991) Breeding of a sake yeast with improved ethyl caproate productivity. *Agric Biol Chem*, **55**, 2153–2154.

48. Fukuda, K., Watanabe, M., Asano, K., Ueda, H., and Ohta, S. (1990) Breeding of brewing yeast producing a large amount of beta-phenylethyl alcohol and beta-pheylethyl acetate. *Agric Biol Chem*, **54**, 269–271.

49. Running, J., Huss, R., and Olson, P. (1994) Heterotrophic production of ascorbic acid by microalgae. *J Appl Phycol*, **6**, 99–104.

50. Backus, M.P. and Stauffer, J.F. (1955) The production and selection of a family of strains in Penicillium chrysogenum. *Mycologia*, **47**, 429–463.

51. Peberdy, J.F. (1985) *Biology of Penicillins*. Benjamin-Cummings, Menlo Park.

52. Angelov, A., Karadjov, G., and Roshkova, Z. (1996) Strains selection of baker's yeast with improved technological properties. *Food Res Int*, **29**, 235–239.

53. Muller, H. (1964) The relation of recombination to mutational advance. *Mutat Res*, **1**, 2–9.

54. Crow, J.F. and Felsenstein, J. (1968) The effect of assortative mating on the genetic composition of a population. *Eugen Q*, **15**, 85–97.

55. Peberdy, J.F. (1979) Fungal protoplasts: isolation, reversion and fusion. *Annu Rev Microbiol*, **33**, 21–39.

56. Seki, T., Choi, E.H., and Ryu, D. (1985) Construction of killer wine yeast strain. *Appl Environ Microbiol*, **49**, 1211–1215.

57. Bortol, A., Nudel, C., Fraile, E., de Torres, R., Giulietti, A., Spencer, J., and Spencer, D. (1986) Isolation of yeast with killer activity and its breeding with an

industrial baking strain by protoplast fusion. *Appl Microbiol Biotechnol*, **24**, 414–416.

58. Kauffman, S. (1993) *The Origins of Order*. Oxford University Press, Oxford.

59. Davies, O. (1964) Screening for improved mutants in antibiotic research. *Biometrics*, **20**, 576–591.

60. Ditchburn, P., Giddings, B., and Macdonald, K.D. (1974) Rapid screening for isolation of mutants of Apergillus nidulans with increased penicillin yields. *J Appl Bacteriol*, **37**, 515–523.

61. Chang, L.T. and Elander, R.P. (1979) Rational selection for improved cephalosporin C productivity in strains of *Acremonium chrysogenum* Gams. *Dev Ind Microbiol*, **20**, 367–379.

62. Rowlands, R.T. (1984) Industrial strain improvement: rational screens and genetic recombination techniques. *Enzyme Microb Technol*, **6**, 290–300.

63. Unowsky, J. and Hoppe, D.C. (1978) Increased production of antibiotic aurodox (X-5108) by aurodox-resistant mutants. *J Antibiot*, **31**, 662–666.

64. Tyo, K.E., Zhou, H., and Stephanopoulos, G.N. (2006) High-throughput screen for poly-3-hydroxybutyrate in *Escherichia coli* and Synechocystis sp strain PCC6803. *Appl Environ Microbiol*, **72**, 3412–3417.

65. Grigg, G. (1952) Back mutation assay method in micro-organisms. *Nature*, **169**, 98–100.

66. Scott, B.R., Alderson, T., and Papworth, D.G. (1973) Effect of plating densities on retrieval of methionine suppressor mutations after ultraviolet or gamma irradiation of Aspergillus. *J Gen Microbiol*, **75**, 235–239.

67. Trilli, A., Michelini, V., Mantovani, V., and Pirt, S.J. (1978) Development of agar disk method for rapid selection of cephalosporin producers with improved yields. *Antimicrob Agents Chemother*, **13**, 7–13.

68. Rous, C.V., Snow, R., and Kunkee, R.E. (1983) Reduction of higher alcohols by fermentation with a leucine-auxotrophic mutant of wine yeast. *J Inst Brew*, **89**, 274–278.

69. Yomano, L.P., York, S.W., and Ingram, L.O. (1998) Isolation and characterization of ethanol-tolerant mutants of *Escherichia coli* KO11 for fuel ethanol production. *J Ind Microbiol Biotechnol*, **20**, 132–138.

70. Sauer, U. (2001) Evolutionary engineering of industrially important microbial phenotypes. *Adv Biochem Eng Biotechnol*, **73**, 129–169.

71. Wisselink, H.W., Toirkens, M.J., Wu, Q., Pronk, J.T., and van Maris, A.J.A. (2009) Novel evolutionary engineering approach for accelerated utilization of glucose, xylose, and arabinose mixtures by engineered *Saccharomyces cerevisiae* strains. *Appl Environ Microbiol*, **75**, 907–914.

72. Helling, R.B., Vargas, C.N., and Adams, J. (1987) Evolution of *Escherichia coli* during growth in a constant environment. *Genetics*, **116**, 349–358.

73. Paquin, C.E. and Adams, J. (1983) Relative fitness can decrease in evolving asexual populations of *S. cerevisiae*. *Nature*, **306**, 368–371.

74. Fleming, G., Dawson, M.T., and Patching, J.W. (1988) The isolation of strains of Bacillus subtilis showing improved plasmid stability characteristics by means of selective chemostat culture. *J Gen Microbiol*, **134**, 2095–2101.

75. Bryson, V. and Szybalski, W. (1952) Microbial selection. *Science*, **116**, 45–51.

76. Butler, P.R., Brown, M., and Oliver, S.G. (1996) Improvement of antibiotic titers from Streptomyces bacteria by interactive continuous selection. *Biotechnol Bioeng*, **49**, 185–196.

77. Naki, D., Paech, C., Ganshaw, G., and Schellenberger, V. (1998) Selection of a subtilisin-hyperproducing Bacillus in a highly structured environment. *Appl Microbiol Biotechnol*, **49**, 290–294.

78. Duetz, W.A., Ruedi, L., Hermann, R., O'Connor, K., Buchs, J., and Witholt, B. (2000) Methods for intense aeration, growth, storage, and replication of bacterial strains in microtiter plates. *Appl Environ Microbiol*, **66**, 2641–2646.

79. Verstrepen, K.J., Chambers, P.J., and Pretorius, I.S. (2006) In Querol, A. and Fleet, G. (eds.), *The Yeast Handbook*. Springer-Verlag, Berlin, pp. 399–444.

80. Hall, B.G., Yokoyama, S., and Calhoun, D.H. (1983) Role of cryptic genes in microbial evolution. *Mol Biol Evol*, **1**, 109–124.

81. Bailey, J.E., Sburlati, A., Hatzimanikatis, V., Lee, K., Renner, W.A., and Tsai, P.S. (2002) Inverse metabolic engineering: a strategy for directed genetic engineering of useful phenotypes. *Biotechnol Bioeng*, **79**, 568–579.

82. Nunokawa, Y. and Yoshizawa, K. (1984) Physiological aspects of yeast in sake fermentation. *Crit Rev Biotechnol*, **2**, 193–231.

83. Kotaka, A., Sahara, H., Kondo, A., Ueda, M., and Hata, Y. (2009) Efficient generation of recessive traits in diploid sake yeast by targeted gene disruption and loss of heterozygosity. *Appl Microbiol Biotechnol*, **82**, 387–395.

84. Katou, T., Kitagaki, H., Akao, T., and Shimoi, H. (2008) Brewing characteristics of haploid strains isolated from sake yeast Kyokai No. 7. *Yeast*, **25**, 799–807.

85. Katou, T., Namise, M., Kitagaki, H., Akao, T., and Shimoi, H. (2009) QTL mapping of sake brewing characteristics of yeast. *J Biosci Bioeng*, **107**, 383–393.

86. Ouchi, K. and Akiyama, H. (1971) Non-foaming mutants of sake yeasts selection by cell agglutination method and by froth flotation method. *Agric Biol Chem*, **35**, 1024–1032.

2

TRACER-BASED ANALYSIS OF METABOLIC FLUX NETWORKS

Michael Dauner

2.0 INTRODUCTION

Biological systems are complex networks consisting of many thousands of components interacting with each other to constitute a phenotype. Currently our knowledge of the cellular parts or of their interactions is not sufficient to model and predict behavior of most biological systems with sufficient accuracy *in silico* as to render experiments *in vivo* unnecessary. It is for this reason, as described in Chapter 1, that classical strain engineering approaches, which entail mutagenesis and screening, were used predominantly to obtain improved traits in industrial strains. However, it is exactly this complexity that calls for engineering approaches based on mathematical models for developing biocatalysts, as mathematical models allow for a systematic integration and analysis of a wealth of information that is currently being generated for many industrial strains (1). The advent of the "omics" technologies (2–6) at the end of last and beginning of the 21st century, providing cell-wide information on genomes (7–9), transcriptomes (10,11), proteomes (12–14), and metabolomes (15–18) further emphasized this need. It also illustrated that knowledge of only the parts of a system does not necessarily translate into more efficient product development, as was illustrated by the inability of the pharmaceutical industry to develop new drugs despite a flood of new data and insights from "omics" analysis (19). This understanding finally gave rise to the renewed interest in the concepts of systems biology and its tools (20,21).

Engineering Complex Phenotypes in Industrial Strains, First Edition. Edited by Ranjan Patnaik.
© 2013 John Wiley & Sons, Inc. Published 2013 by John Wiley & Sons, Inc.

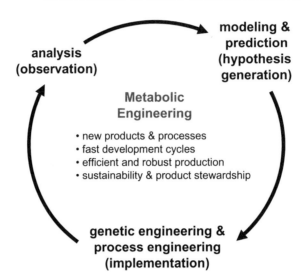

FIGURE 2.1. Metabolic engineering workflow comprising hypothesis generation, implementation, and observation.

Metabolic engineering (22) represents a systems biology approach to the analysis and design of metabolic flux networks (23). Successful metabolic engineering approaches build on a close interaction between hypothesis generation, implementation, and observations and aim to develop new products and processes, to shorten development times, to improve efficiency and robustness of processes, and to ensure sustainability and minimal health and environmental impact of the product and processes over their life time (Figure 2.1). The metabolic engineering toolbox comprises many experimental, analytical, and mathematical techniques and methods, from recombinant DNA technology (24) to directed evolution (25) to sophisticated modeling approaches (26). Among its most prominent tools are frameworks for the analysis and design of metabolic flux networks, in particular network analysis (NA), (stoichiometric) metabolic flux analysis (MFA), their application to dynamic conditions (dynamic MFA, D-MFA) and complex cellular systems ("*in silico* cells"), as well as their recent extension to integrate data from tracer experiments, predominantly employing the nonradioactive carbon isotope ^{13}C (^{13}C NA and ^{13}C MFA), but also other tracers such as, ^{15}N or ^{31}P (Figure 2.2). Fluxes, equivalent to reaction rates, cannot be measured directly. It is only from observing and balancing changes of, for example, substrate or product levels, or other quantities related to a reaction rate, such as temperature or pressure, that fluxes can be inferred from. However, in biotechnological applications, fluxes (rates) generate the desired amounts of product (titer) while producing as few as possible by-products (yield). It is this combination of rate, titer, and yield (RTY) values that form the core

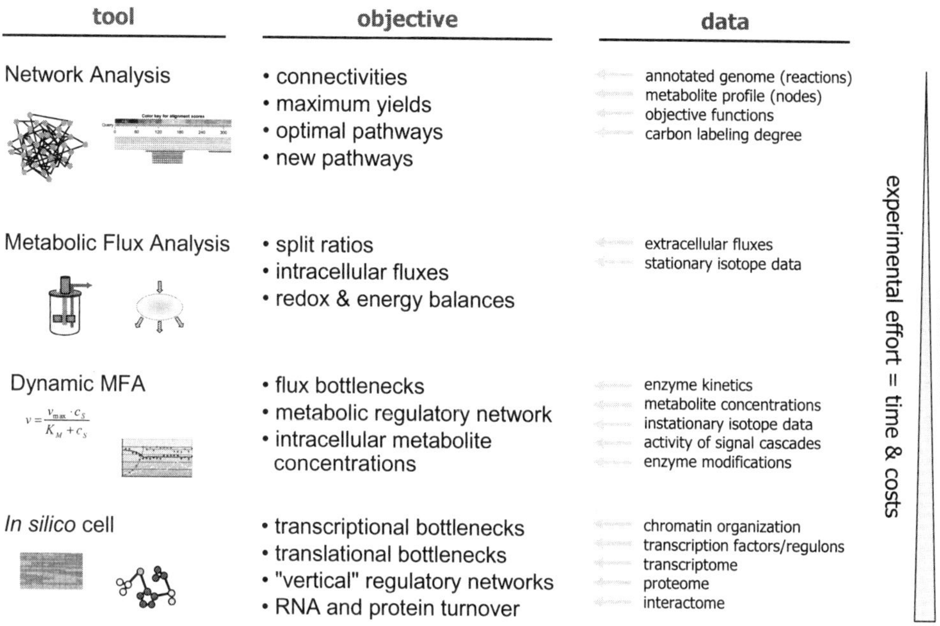

tool	objective	data
Network Analysis	• connectivities • maximum yields • optimal pathways • new pathways	annotated genome (reactions) metabolite profile (nodes) objective functions carbon labeling degree
Metabolic Flux Analysis	• split ratios • intracellular fluxes • redox & energy balances	extracellular fluxes stationary isotope data
Dynamic MFA $v = \frac{v_{mm} \cdot c_S}{K_M + c_S}$	• flux bottlenecks • metabolic regulatory network • intracellular metabolite concentrations	enzyme kinetics metabolite concentrations instationary isotope data activity of signal cascades enzyme modifications
In silico cell	• transcriptional bottlenecks • translational bottlenecks • "vertical" regulatory networks • RNA and protein turnover	chromatin organization transcription factors/regulons transcriptome proteome interactome

experimental effort = time & costs

information

FIGURE 2.2. Metabolic engineering tools for the analysis of metabolic fluxes. *(See insert for color representation of the figure.)*

performance metrics in the development of almost all new biocatalysts and bioprocesses.

NA and MFA and their extension to dynamic and whole-cell systems are true engineering approaches to design and balance flux networks using model-based analyses that take advantage of the conservation laws of elements, mass, and recently with the incorporation of thermodynamic approximations, the conservation of energy. The models consisting of a set of balance equations help not only in designing optimum production pathways, but also in structuring complex data sets and in superimposing additional constraints that allow for deriving unique insights into complex networks of biological systems by noninvasive means. In case redundant information is available, consistency of measurements can be evaluated. This chapter, along with Chapter 3, is devoted to illustrating the basic steps in designing and analyzing new biotechnological processes and production systems with the help of NA and MFA. However, the emphasis of this chapter will be on elucidating the advantages as well as drawbacks of integrating tracer data into NA and MFA, while Chapter 3 will focus on the integration of "omics" data into NA and MFA. In addition to detailing how to carry out a tracer experiment and analyze isotope data, examples are presented in detail where tracer-based NA and MFA were successfully applied to facilitate engineering of new biotechnological processes and products.

2.1 SETTING UP A STOICHIOMETRIC NETWORK MODEL

The first step in NA and MFA is to set up a representative stoichiometric model comprising all relevant reactions of the system. In metabolic networks, fluxes usually rely on enzyme catalysis. Consequently knowledge of all catalytic genes in an organism and the stoichiometry of the reaction(s) they catalyze allows for the reconstruction of the metabolic network from the annotated genome (27). With sequencing capabilities improving at a rapid pace, the number of available sequenced genomes has increased dramatically. The first genome sequence of *Haemophilus influenza* was released in 1995 (7). Today genome sequences of more than 180 organisms have already been completed (28).

However, identification of all open reading frames (ORFs) in an organism is a challenge (29), and even in the well-characterized model organisms *Escherichia coli* and *Saccharomyces cerevisiae*, 940 out of 4472 (21%) and 1134 out of 5796 (20%) (30) protein coding genes are still uncharacterized, respectively. Also, nonenzymatic reactions that occur under physiological conditions need to be integrated into the model. Examples are the hydrolysis of phosphogluconolactonate (31) in the oxidative pentose phosphate pathway (PPP), the decomposition of acetolactate to acetoin or diacetyl (32) in the branched-chain amino acid biosynthesis pathway, or of glutamine to ammonia and pyrrolidonecarboxylic acid (33). But simple chemical reactions under physiological conditions also play a role in regulatory processes, for example, in the formation of nitric oxide, an important cellular signaling molecule, from hydrogen peroxide and arginine (34), or in the reactions of 2,4-dienone 13-oxooctadecadienoic acid, a regulator in several cellular processes, with glutathione and N-acetylcysteine (35).

Another challenge in setting up a network model results from promiscuous enzymes (36,37). Last but not least, often stoichiometry of a reaction is not known. One of the most prominent examples is the adenosine triphosphate (ATP) synthetase reaction, coupling ATP generation with flux of protons over the membrane. A general H^+-to-ATP ratio of 4 was assumed (38). However, recently insights into the molecular mechanism of the enzyme and its bioenergetics point to flexible H^+-to-ATP ratios, depending on the structure and localization of the ATP synthetase in the respective organism (39). To further complicate matters, frequently also stoichiometry of H^+ transport by the respiration chain is difficult to determine, not only because of differing constituents of the respiratory chain (40), but also because of electron slippage (41), parallel reactions such as alternative oxidases and uncoupling proteins (42), or proton and electron leakage (43). There are many more examples of physiology-dependent reaction stoichiometries, including ion leakage through membranes (44) or different cofactor preferences of the glucose-6-phosphate dehydrogenase (45). Therefore, care needs to be taken to keep experimental conditions defined and reproducible, as well as to have a detailed understanding of these mechanisms in the specific system under investigation.

Instead of composing a metabolic network model based on its known reactions, alternatively an indirect approach can be taken starting from its known metabolites. With recent progress in comprehensive metabolite analysis, on a cellular level referred to as metabolomics, detection of a broad set of metabolites is possible (15,46). Combined with knowledge of reaction biochemistry, most of the reaction steps that link the detected metabolites can be derived. However, there may also be cases where parallel reactions with different reaction orders and biochemistry would be possible to explain the detected metabolites. In this case perturbation experiments of the metabolite pools in question would be required, for example, by stimulus-response experiments (47). From observation of the response of the downstream metabolite pools in question, connectivity could be derived. The feasibility of this approach with respect to current experimental capabilities is still limited. Nevertheless, the potential of this alternative avenue to reconstruct a metabolic network is demonstrated by the finding that variations in metabolite levels observed at different steady-state conditions revealed connectivity in the underlying metabolic network (48).

But what if the desired product is a non-natural compound and no biosynthesis pathways for its production are known? Traditionally the knowledge of domain experts on enzymatic reactions and substrate specificities of enzymes is required to explore options for new biosynthesis routes. However, recently a computational framework was developed that allows for a systematic identification of possible reaction pathways from a given set of enzyme reaction rules (49). Molecules are represented using bond–electron matrices (50). Enzyme-catalyzed reactions use a similar notation. The reactive sites for each enzyme class are predefined as two-dimensional (2D) molecule fragments. A set of molecules is given as input and evaluated to determine if it contains compounds with suited functionality to undergo reactions corresponding to the specified reaction classes. The reactions are then implemented through matrix addition (49). This approach opens ways to construct and explore unknown pathways or compare their efficiency with known biosynthesis pathways, as, for example, described for the biosynthesis of 3-hydroxypropanoic acid (51) or 1,4-butanediol (52). Nevertheless, optimum product pathways identified by this method may comprise one or more generic reaction steps for which no enzyme is known to exist, requiring protein engineering efforts to derive the desired activity from homologous enzymes (53). However, until protein design methods improve, uncertainty remains as to whether an enzyme with the required performance can be successfully engineered. Recently dramatic progress was made in computational protein design (54), resulting, for example, in the *de novo* implementation of new reaction chemistry into enzymes, such as a Kemp elimination (55) or a bimolecular Diels–Alder reaction (56). It can be assumed therefore that the reaction space defined by known enzymatic reactions will not be limiting the development of new processes and products in the future, but will be expanded to comprise all feasible and thermodynamically favored chemical reactions.

2.2 SMALL-SCALE MODELS VERSUS GENOME SCALE MODELS

In setting up stoichiometric metabolic models, the metabolic engineer frequently has to make a decision between focused small-scale models and comprehensive large-scale models (Figure 2.3). The advantage of small-scale models is their mathematical amenability and the usually straightforward interpretation of the results obtained. Their construction is "bottom-up," adding one reaction at the time to the model. However, while this "step by step" setup of the model assures a detailed understanding of the model by the metabolic engineer, at the same time it also represents one of the major challenges of working with small-scale models. Decisions need to be made as to which reactions to include or not to include in order not to limit the capabilities of the model to provide mechanistic and quantitative analyses and predictions (57). Frequently the models only comprise reactions of the major catabolic or anabolic pathways in central carbon metabolism, for example, the Emden–Meyerhoff pathway (EMP), the PPP, the tricarboxylic acid cycle (TCA), and, if applicable, the Entner–Doudoroff pathway (EDP) and Calvin–Benson–Bassham cycle (CBB). If industrial processes are analyzed, the production pathway of interest is often included in the model as well. Further model simplification can be achieved by lumping reactions of a linear reaction sequence into one overall reaction (58–63). If the system is growing, consumption of precursors for biomass formation has to be considered in the mass balances of the model as well. For this purpose knowledge on the composition of a cell is needed or at least needs to be approximated, that is, its protein, lipid, carbohydrate, and nucleic acid content. An additional biomass fraction usually

--→

FIGURE 2.3. Genome-scale (A) and small-scale model (B) of *E. coli*. The genome-scale model (reproduced by M. Dauner from a screenshot of a genome-scale model of *E. coli* in Insilico Discovery by J.W. Schmid, 2012, with permission from Insilico Biotechnology (Stuttgart, Germany, http://www.insilico-biotechnology.com)) comprises 849 metabolites and 1334 transformers. A detail of the model as represented in the modeling and simulation environment Insilico Discovery is represented in the right window. Metabolites are displayed as blue and gray circles, metabolic reactions as red, transmembrane transport processes as yellow squares. Prominent pathways are marked by gray boxes. Associated data on transcript, protein, and enzyme levels are accessible in the upper left window, model information in the lower left window. The modeling environment supports stoichiometric as well as dynamic simulations. In addition, effector kinetics can be visualized (data not shown). The small-scale model consists of 23 metabolites and 26 reactions (165). Gray arrows represent fluxes of respective metabolites for biomass formation. Abbreviations: G6P, glucose-6-phosphate; F6P, fructose-6-phosphate; T3P, glyceraldehyde-3-phosphate; P5P, ribose-5-phosphate; S7P, sedoheptulose-7-phosphate; E4P, erythrose-4-phosphate; PGA, 3-phosphoglycerate; SER, serine; GLY, glycine; C1, methyl group bound to tetrahydrofolate; PEP, phosphoenolpyruvate; PYR, pyruvate; ACA, acetyl-CoA; MAL, malate; FUM, fumarate; OGA, α-ketoglutarate; OAA, oxaloacetate; TCA, tricarboxylic acid. *(See insert for color representation of the figure.)*

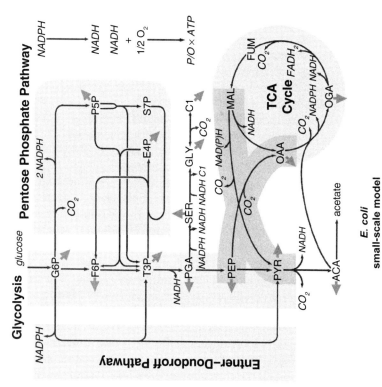

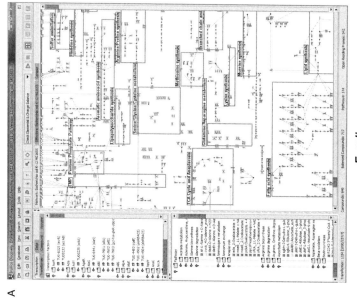

A

E. coli
genome-scale model

B

comprises amounts of intracellular low molecular weight compounds such as vitamins, pathway intermediates, and salts. In case of macromolecules, for example, proteins or DNA, further information on their monomer composition is required, that is, their amino acid and nucleic acid fractions, respectively (59–62).

In small-scale models, assumptions on the biosynthetic pathways of these biomass precursors are made that allow for finally expressing biomass biosynthesis as a lumped reaction comprising only input metabolites with a single "biomass" variable as output (along with some other metabolites produced, e.g., NADH, CO_2, and ADP). However, it is this less stringent and, with respect to the model balance equations, "hidden" process of deriving a lumped reaction of biomass synthesis that represents one of the major differences in the use of small-scale versus genome-scale models. In genome-scale models (see Chapter 3) no reaction lumping and assumptions on biosynthesis pathways is necessary. Model construction is "top down," which means it starts with the list of all reactions expected to occur in the cell. Valuable insights can be gained into complex biosynthesis requirements and pathways that can function as bypass reactions and that would usually not be considered in the small-scale model (64). Nonetheless, genome-derived metabolic networks typically contain a large set of missing reaction steps and dead-end metabolic pathways to be curated (65). As a result, a large degree of freedom due to numerous parallel pathways and redundant reactions frequently render analysis of the model tedious. Dealing with reaction directionality in a large set of metabolic reactions can be a time-consuming task if thermodynamic constraints are not already integrated into the model (66).

2.3   NETWORK ANALYSIS: MAXIMUM THEORETICAL YIELD

Once a stoichiometric network model of an organism is set up, a frequently encountered question in industrial biotechnology is the maximum yield of a product that is achievable with a given substrate in the respective organism. With either a small-scale or genome-scale stoichiometric network model at hand, this question translates mathematically into a linear optimization problem: given a certain input into a network, what is the maximum output of the desired compound? In industry this analysis is routinely carried out as a first step in the evaluation of the economic feasibility of a new product or process idea, the so-called techno-economic assessment. Results of such a theoretical yield analysis are, for example, our knowledge on the maximum theoretical yield of riboflavin that can be achieved with *Bacillus subtilis* on glucose: 0.257 mol/mol (59), of penicillin V with *Penicillium chrysogenum*: 0.43 mol/mol (67), of 1,3-propanediol with *E. coli*: 1.49 mol/mol (68); and so on. As costs of the raw materials are known, the maximum theoretical yield allows for deriving the minimum material costs required to produce a product in the best thinkable process. The calculation also incorporates information on co-substrates that need to be provided (e.g., oxygen), as well as waste products that are produced (e.g., CO_2).

Another outcome of the analysis is knowledge of the optimum metabolic path associated with the maximum theoretical yield. This knowledge can be used in developing and evaluating the feasibility of metabolic engineering strategies for biocatalyst construction. The more reaction steps a production pathway comprises, not only the more complicated its implementation will be, for example, with respect to the genetic engineering required, but also the higher is the expected loss of carbon due to side reactions. This in turn results in the anticipation of an increased effort necessary for its optimization (69,70).

In cases where the network contains parallel reactions or pathways with similar cofactor requirements, no unique optimum pathway solution exists. Linear optimization will only deliver one of the solutions but will fail to identify all equivalent routes. An alternative to address this shortcoming of linear optimization is to apply "elementary mode analysis." In elementary mode analysis all feasible "steady-state" modes to start from a substrate and to yield a product that at the same time cannot be simplified (decomposed) further are identified (71). The solutions can be ranked according to their yields, with the highest yield solution representing the maximum theoretical yield. In the case of networks with parallel reactions or pathways with similar cofactor requirements, elementary mode analysis will identify all possible routes and will present them as solutions with equal maximum theoretical yield. Finally, all possible pathway solutions yielding maximum theoretical yield can be represented by linear combinations of this set of optimum elementary flux modes. Studies of elementary mode analysis for optimum production modes were, for example, carried out for lysine and *Corynebacterium glutamicum* (72), the metabolic engineering of *Saccharomyces cerevisiae* for poly-β-hydroxybutyrate formation (73), or the production of succinate by *E. coli* (74). However, while elementary mode analysis provides a more comprehensive understanding of the system than linear optimization methods, this advantage comes at the costs of significantly increased computational effort (75). This chapter does not provide room for an in-depth discussion of the advantages and limitations of "elementary mode analysis." Readers interested in more details are referred to Reference (76).

2.4 (STOICHIOMETRIC) METABOLIC FLUX ANALYSIS

While NA only uses the stoichiometry of the reaction network, stoichiometric MFA, frequently also included in the term flux balance analysis (FBA) (77), additionally introduces a time aspect, usually in the form of specific fluxes of a substrate or product i, q_i. Specific fluxes can be derived from measured volumetric reaction rates (Q_i), usually determined in "g/L/h" or "mmol/L/h," by normalizing them to the actual cell concentration cx, typically determined as grams of "cell dry weight" per liter ("g (cdw)/L"), resulting in the units of either "g (i)/g (cdw)/h" or "mmol (i)/g (cdw)/h." However, from practical considerations, molar units are preferred to mass units, as reaction stoichiometries usually refer to molar units. Remarkably, apart from the specific product and substrate fluxes, another commonly used specific flux is the specific growth rate mu, defined as mu = $1/cx \, dcx/dt$, which essentially represents the specific flux of biomass formation. Specific fluxes provide a means of directly assessing cellular physiology/biocatalyst performance independent from process variables such as the biomass/biocatalyst concentration. They are of utmost importance in determining progress in strain development efforts in the engineering of complex phenotypes.

A common misconception is that MFA can only be applied under steady-state conditions, that is, in conditions where intracellular and extracellular metabolite and biomass concentrations are not changing, as, for example, encountered in steady-state chemostat cultivations. However, MFA can be applied as well to quasi-steady state, yet even highly dynamic conditions, as long as the mass balances applied hold for the respective time interval. In this case MFA yields the average specific fluxes for the analyzed time frame. However, care needs to be taken such that in order to obtain physiologically meaningful results the cell concentration used for normalization under dynamic conditions is not a constant but represents the average cell concentration cx, corresponding to the integral of the cell concentration from the start to the end of the investigated time interval Δt. In a batch culture growing exponentially with the maximum specific growth rate mu_{max}, the average cell concentration can, for example, be calculated according to the following equation: $cx_{ave} = (cx_{end} - cx_{start})/(\Delta t \times mu_{max})$.

The stoichiometric matrix used in NA holds the information on the network topology. Every row represents a mass balance and every column the stoichiometric coefficient of a reaction in the respective mass balance. In stoichiometric MFA the stoichiometric matrix is multiplied with, for example, (molar) specific fluxes to yield a homogenous linear equation system. If some of these fluxes, typically extracellular fluxes, are determined experimentally, the system can be further decomposed to eliminate these known fluxes. However, if the number of the remaining independent mass balances (or mathematically speaking, the rank of the matrix) is smaller than the number of unknown fluxes considered in the analysis, the system is underdetermined. This means an unlimited number of solutions that can be represented by a linear combination of a set of independent fluxes exist. If the number of independent mass balances is equal to the number of unknown reactions, the system is determined and can be solved to yield either none or a unique flux solution. However, the system can also be overdetermined with more mass balance equations and measured fluxes than unknown fluxes. In this case the redundant measurement information can be used for measurement data reconciliation. For more information on the mathematical details of stoichiometric MFA the interested reader is referred to Reference (78). While mass balances on metabolite pools are most broadly used, all other entities that can be balanced can also be integrated into the equation system, for example, elemental (79) or heat balances (80).

Stoichiometric MFA is a remarkable tool in that it allows the metabolic engineer to noninvasively assess intracellular flux distributions based on measurements from outside the cell. The tool is particularly useful for investigating redox and energy metabolism, because, for example, redox metabolites such as nicotinamide adenine dinucleotide (NAD(H)) and nicotinamide adenine dinucleotide phosphate (NADP(H)) and energy metabolites such as adenosine mono-, di-, and triphosphates (AMP, ADP, ATP) participate in many different reactions in metabolism and their metabolite pools exhibit very high turnover rates. As discussed previously, a resolution of the intracellular fluxes is only possible if the resulting equation system is determined (81). However, due to duplicate reactions and parallel pathways, and because not all of the ATP production and consumption processes can be quantitatively accounted for, the resulting equation system is usually under- rather than (over-)determined. Additional balances that can be incorporated into the analysis (82) are required. Balances on isotope provide such additional information.

2.5 CARRYING OUT A LABELING EXPERIMENT

Frequently labeling experiments are carried out while metabolism of the organism is in "steady state," also referred to as "stationary." During steady state, intracellular concentrations of metabolites do not change. Consequently metabolite and label balances can be drawn on the basis of constant fluxes and accumulation of metabolites does not need to be considered. Steady-state conditions are best achieved in continuous culture experiments, where the specific growth rate of the culture is determined by the dilution rate. Another advantage of chemostat experiments is that the biomass concentration of the culture can be set by the concentration of the limiting substrate in the feed. As comprehensive analysis of labeling experiments frequently requires large amounts of samples, for example, if fluxes are analyzed in combination with the transcriptome and proteome, sufficient amounts of samples can be generated. Usually the labeling experiment is started the moment the investigated culture reaches steady state. As a rule of thumb, it will take 3–5 volume changes after the onset of substrate limitation to reach steady state, corresponding to approximately 95% or 99% of cells being newly generated under substrate limiting conditions, respectively. During the labeling experiment, unlabeled substrate in the feed (of course usually still containing the respective isotope at natural abundance) is replaced with a substrate specifically labeled with an isotope in certain positions. Most frequently used isotope substrates are glucose labeled with ^{13}C in position C1 ($^{13}C_1$-glucose), glucose uniformly labeled in all positions (u-^{13}C glucose), or ammonium labeled with ^{15}N (^{15}N-ammonium).

In order to describe positional labeling patterns of isotopes in molecules, they are frequently referred to as "isotopomers." The term "isotopomer" is an acronym and stands for "isotope isomer." A metabolite with n atoms of an element and m isotopes has m^n isotopomers, corresponding to all possible positional combinations of the isotopes in the molecule. Pyruvic acid, for example, has three carbon atoms and eight isotopomers of the carbon isotopes ^{12}C and ^{13}C. Optimal isotopomer composition of the substrate(s) for a labeling experiment is best determined based on the focus of the analysis, availability of labeled substrates, and hypothesized flux distribution in the system, applying optimum experimental design (83). With the replacement of the substrate with its labeled mixture, the culture is still in a metabolic steady state but is instationary with respect to the distribution of its labeling patterns in metabolites and macromolecules.

Usually pool sizes of intracellular metabolites are small compared with the fluxes, so the resulting high turnover rates result in an isotopic steady state of the metabolite pools rather quickly. However, the advantage of using building blocks of macromolecules rather than metabolites for label analysis is that no fast sampling methods need to be applied, as turnover times of macromolecular pools are rather large. In particular, amino acids that can be gained from cellular protein by hydrolysis allow for broad insights at various points of central carbon metabolism (Figure 2.4). If building blocks in macromolecules rather than metabolites are used for analysis, it takes 3–5 volume changes to get close to an isotopic steady state, which requires considerable amounts of labeled substrates for the experiment. Nevertheless, isotopic steady state of the macromolecular labeling composition is not a *conditio sine qua non*, but can be approximated by assuming simple washout kinetic for all macromolecules generated before the onset of the labeling experiment according to fr (unlabeled) = exp $(-D \times t_{exp})$, with fr (unlabeled) representing the fraction of unlabeled macromolecule, D the dilution rate of the continuous culture, and t_{exp} the duration of the labeling experiment. Wiechert et al. (84) approximated the difference between the two time constants governing the turnover of metabolite pools and the turnover of macromolecules to be three orders of magnitude, that is, 0.01 hour and 10 hours, respectively. These approximations assume that no dilution of intermediates results from the turnover of macromolecules. Also, specific fluxes into a pool are presumed to be in the order of the specific molar glucose uptake. However, if the metabolic flux network is significantly engineered, labeling patterns in intermediates may not be representative. This was for example shown for *C. glutamicum*, where the deletion of pyruvate dehydrogenase resulted in very low TCA fluxes (85). Moreover, macromolecules such as proteins (86) are constantly assembled from and again decomposed into their monomers, which results in label dilution of intermediates that can significantly bias the outcome of labeling experiments (87). Nevertheless, the labeling information of proteinogenic amino acids can also be used to correct measured mass isotopomer distributions of free amino acids for any dilution effects, as, for example, reported by Iwatani et al. (88).

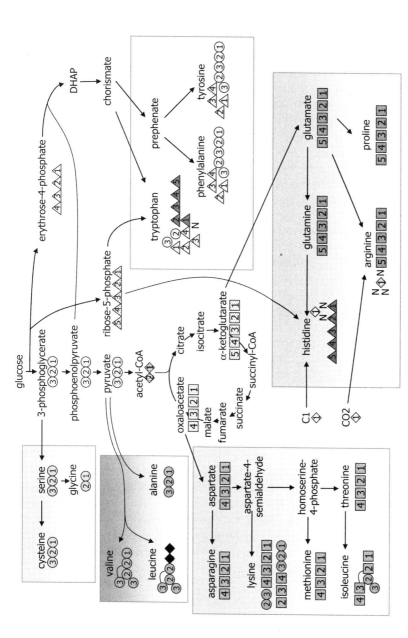

FIGURE 2.4. Origin of carbon atoms in amino acids from precursor metabolites in central carbon metabolism. Green box: serine family. Maroon box: pyruvate family. Purple box: aspartate family. Mocca box: glutamate family. Yellow box: shikimate family. DHAP, 3-deoxy-D-arabino-heptulosonate-7-phosphate; CoA, coenzyme A. (*See insert for color representation of the figure.*)

If a protein is produced specifically by an organism, it can be purified and used to conclude on split ratios of fluxes of the particular organism, even if this organism was growing in a mixed culture (89). However, it will still be a challenge to combine the derived split ratios with extracellular fluxes, as these fluxes can hardly be measured separately.

A process regime frequently used in biotechnology is batch cultures. In batch cultures all the substrates are provided at the beginning of the process. No additional substrates are fed nor is any culture broth harvested. Under this condition the microorganism does not experience any substrate limitation and usually grows (substrate inhibitions not considered) at its μ_{max} in the respective medium until either a substrate becomes limiting or an extracellular by-product reaches inhibiting levels. The system is metabolically in "pseudo" steady state or "quasi" stationary, as intracellular fluxes and concentrations are virtually constant. This setting allows for carrying out meaningful flux and tracer experiments in batch cultures (90), making also small-scale experiments in microtiter or deep-well cultures—cultivation platforms frequently used in high-throughput screening (see Chapter 5)—possible (91,92).

Recording the degree and pace of label distribution in metabolites in a metabolically stationary phase provides valuable information for so-called isotopically instationary MFA. Analysis of the isotopomer time profiles of metabolites enables determination of fluxes with improved accuracy, and can also be used for the prediction of intracellular concentrations of metabolites that cannot be measured directly (93–96). However, systems boundaries in analysis need to be selected with care as turnover of large intracellular pools such as storage carbohydrates in *S. cerevisiae* can significantly influence isotope labeling patterns of the analyzed metabolites, similar to the metabolic and isotopic stationary case (97).

Isotopically instationary conditions are also encountered in the analysis of industrially relevant fed-batch cultures. If labeling information from proteinogenic amino acids is exploited, high biomass concentrations, slow growth rates, and high fractions of substrate consumed for maintenance metabolism require the use of large amounts of labeled substrates, as was the case in an industrial relevant fed-batch process of a recombinant *E. coli* producing 1,3-propanediol (98). However, if metabolites such as free amino acids are used, the amount of labeled substrates can be significantly reduced (99).

Last but not least there are stimulus–response experiments (SREs). SREs were established to assess fast regulatory loops on the metabolite level *in vivo* (100). Cells are grown under (quasi) steady-state conditions and perturbed by a sudden external stimulus. Rapid sampling and quenching procedures are required, as the flux and metabolite network responds within a sub-second time scale (101,102). By this approach fast processes at the level of metabolite regulation can be investigated without interference from the slow gene regulatory processes (103). However, in applying several simplified model descriptions, it was found that the perturbations in metabolite concentrations of a single SRE were too small to allow for a complete elucidation of the investigated metabolic flux system (104). Significantly more

	MFA			Dynamic MFA	
metabolic state	stationary			instationary	
isotopic state	no label	stationary	instationary	no label	instationary
measurements	• concentrations of extracellular metabolites	• concentrations of extracellular metabolites • labeling patterns of macromolecule building blocks	• concentrations of extracellular metabolites • labeling patterns of intracellular metabolites	• concentrations of extracellular & intracellular metabolites	• concentrations of extracellular & intracellular metabolites • labeling patterns of extracellular & intracellular metabolites
variables	• net fluxes	• net & exchange fluxes • split ratios	• intracellular metabolite concentrations • net & exchange fluxes • split ratios	• enzyme kinetics • intracellular metabolite concentrations • net fluxes = $f(t)$	• enzyme kinetics • intracellular metabolite concentrations • net & exchange fluxes = $f(t)$ • split ratios = $f(t)$
information	+	++	+++	++	++++
complexity					

FIGURE 2.5. Classification of experiments at various combinations of metabolic and isotopic state together with an overview of the minimum associated measurements, variables that are required to describe the experiment, as well as the amount of information and complexity that need to be handled.

information can be obtained if the experiment is augmented with isotopic tracer data, called D-MFA (105). An overview of the different combinations of metabolic and isotopic stationary as well as instationary experiments is given in Figure 2.5.

2.6 MEASURING ISOTOPE LABELING PATTERNS

Isotopes are variants of atoms of a particular chemical element that have differing numbers of neutrons and consequently differ in their mass, for example, ^{13}C or ^{15}N. This difference in mass can be detected by mass spectrometry (MS) (Figure 2.6). The name "mass spectrometry" is a misnomer as the mass is not what is measured. Instead, MS determines the mass-to-charge (m/z) ratio or a property related to m/z. A mass spectrum is a plot of ion abundance versus m/z, although in many cases the x-axis is labeled "mass" rather than m/z (106). Frequently analysis of complex sample mixtures requires the combination of both separation techniques and MS. For volatile compounds, separation by gas chromatography (GC) is frequently applied. For highly polar nonvolatile compounds, separation of liquid chromatography (LC) is the method of choice. The hyphen used to indicate the coupling of a separation technique to MS, for example, GC-MS or LC-MS, led to the group term "hyphenated methods" (107). An obvious limitation of hyphenated methods based on GC or LC

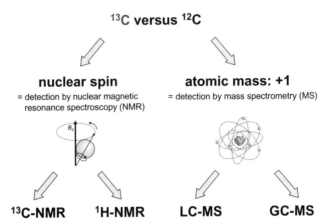

FIGURE 2.6. Measurement methods for the detection and quantification of ^{13}C labeling patterns in metabolites.

technologies is that the sample path is also acting as a filter and not all components injected will necessarily pass through. Components will therefore remain in the injector, column, and detector and the whole system will be inherently different after each injection. This is in contrast to nuclear magnetic resonance (NMR), which is a nondestructive spectroscopic technique (108). NMR is an alternative for analyzing isotope labeling patterns. However, only isotopes with an odd number of protons and/or of neutrons have an intrinsic magnetic moment and angular momentum, in other words a nonzero spin, and can therefore be detected by NMR. Nuclei frequently studied by NMR in biology are ^{1}H, ^{13}C, ^{15}N, and ^{31}P (109) (Figure 2.6).

In GC, sample mixtures are separated based on gaseous–solid phase interactions. Briefly, the sample solution is injected into a heated sample port and is vaporized. A carrier gas, most of the time helium or nitrogen, sweeps the vaporized sample molecules into a column. The velocity at which a compound transitions through the column depends on the strength of its adsorption, which in turn depends again on its molecular structure, the stationary phase material, and the temperature. The column is located in an oven to control the temperature according to a program. Separation of the sample mixture is therefore primarily based on boiling point and vapor pressure differences between its components, and to a lesser extent on their interactions with the stationary phase of the column. Carrier gas flow and column properties such as coating, diameter, and film thickness, in combination with the temperature program, result in a characteristic elution profile and retention time of each molecule. For a more detailed description of GC principles the reader is referred to Reference (110). GC works well for volatile compounds with boiling points below 300°C. Most of the commonly used columns are not suitable for operation at significantly higher temperatures, due to limited thermal stability of their stationary phases. However, the majority of metabolites relevant in biotechnological applications are either polar compounds with low vapor pressure and significantly higher boiling points, or not stable at these temperatures, for example, most sugars and amino acids. The boiling point of a molecule usually increases with its molecular weight and number and polarity of its functional groups. These polar groups allow for dipole–dipole interactions and the formation of hydrogen bonds between the molecules and sample matrix. To analyze these compounds by GC, derivatization methods were developed that chemically transform them into compounds with lower boiling points. Organic acids and amino groups are commonly silylated, predominantly with trimethylsilyl (TMS) or tertbutyldimethylsilyl (TBDMS) groups (111). In addition, keto- (oxo-) groups are usually oximated (112) in order to improve their GC properties and prevent enolization reactions, which can introduce multiple products, thereby complicating the chromatograms. However, during high-resolution chromatography, the syn- and anti-isomers of the oximes can sometimes partially separate, giving rise to recognizable shoulders on the GC peaks (108). Alternative derivatization methods were reported; for example, with amino acids the use of (N,N)-dimethylformamide dimethyl acetal gave dimethylaminomethylene methyl esters (113), or derivatization with ethyl chloroformate yielded N-ethoxycarbonyl ethyl esters (114). As the number of derivatized groups increases, there is a danger that the molecular mass of the derivative will be outside the mass range of the detector, typically m/z 650–1000, or will be too high that the derivative will not pass through the GC column. In addition, the likelihood of sterically hindered groups can lead to the formation of multiple products, thereby complicating the chromatogram. There is a wide range of mass spectrometers available varying in the type of ionization and the mass separation. Single quadrupole mass spectrometers with electron impact (EI) ionization are the most often used type of instrument. Compared with other instruments they are relatively low-cost and offer a range of advantages such as high robustness, high sensitivity, and high accuracy of the measured labeling patterns (115). An example flow chart for the analysis of labeling patterns in an amino acid mixture is shown in Figure 2.7.

In LC-MS, for separation, the sample compound mixture is dissolved in a fluid called the "mobile phase," which carries it through a structure holding another material called the "stationary phase." The various constituents of the mixture travel at different speeds, causing them to separate. Many separation modes exploring various interactions are available, for example, reversed-phase, hydrophobic interaction, normal-phase, hydrophilic interaction, ion-exchange, ion-pair, size-exclusion, chiral, ligand exchange, or complexation chromatography (116). After separation, the sample compounds dissolved in the "mobile phase" have to be volatized and ionized to carry out MS. Atmospheric pressure ionization (API)-based interfaces are the most broadly used today, although many other interfaces are available, such as those based on particle-beam, continuous-flow fast atom bombardment, or thermospray

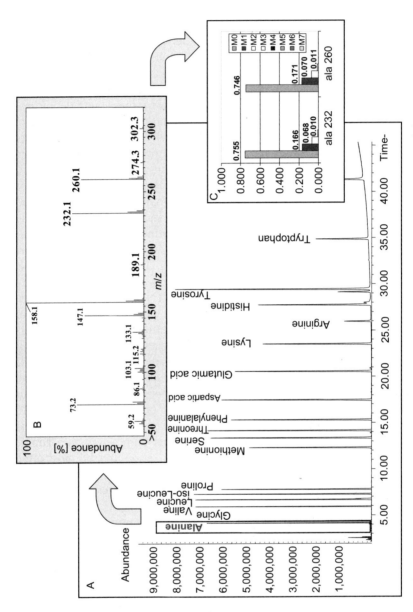

FIGURE 2.7. Chromatogram of TBDMS-derivatized amino acids separated by GC (A), mass spectrum of TBDMS-alanine (B), and label distribution in the characteristic TBDMS-alanine with M0 signals at $m/z = 232$ and $m/z = 260$, respectively. (*See insert for color representation of the figure.*)

(117). Compared with GC-MS, a major advantage of LC-MS using API has been the avoidance of a requirement to derivatize the samples. API sources include both electrospray ionization (ESI) and atmospheric pressure chemical ionization (APCI). In ESI, a high voltage is used to generate droplets containing multiply charged ions. In APCI, the LC column effluent is nebulized into a heated tube, which vaporizes nearly all of the solvent. The solvent vapor acts as a reagent gas and enters the APCI source, where ions are generated with the help of electrons from a corona discharge source. API generally produces much simpler spectra than EI. Depending on the chemical properties of the compound, the ion source design, the ion source potential, the nature of the matrix, and the solvent composition, mainly pseudo-molecular ions ([M+H]$^+$ or [M-H]$^-$) are produced (108). Analysis of the mass spectral of a pseudo-molecular ion provides considerably less information on the isotopomer distribution of a compound than, for example, the multiple fragments obtained by EI. Moreover, in API it is not always simple to predict whether positive or negative ions will be produced (118). Matrix effects can comprise ionization suppression and ionization enhancement caused by salts and other components that are ionized at the same time (119). Due to these reasons, only limited search libraries of product ion mass spectra are available, hampering the fast identification of unknown peaks obtained in LC-MS analyses.

Label analysis is frequently carried out on monomers derived from hydrolysis of macromolecules such as proteins or DNA, while analysis of labeling patterns in intracellular metabolites was long impeded by their low concentrations and high turnover. However, with improved sampling procedures and the constantly increasing sensitivity of the analytical methods, these obstacles were overcome. Free intracellular amino acids were analyzed by GC-MS by Wittmann et al. (120). LC-MS-MS was used to determine mass isotopomer distribution in free and proteinogenic amino acids (88), 40 mass isotopers of 10 phoshorylated compounds (121), or 60 mass isotopomers from 13 metabolites (122). In addition to eight phosphorylated compounds, mass isotopomers of pyruvic acid, alpha-ketoglutaric acid, succinic acid, glutamic acid, and aspartic acid were analyzed. Coupling capillary electrophoresis with MS allowed the determination of 73 mass isotopomers from 13 metabolites (123). Nevertheless, LC-based separation methods are less sensitive and less accurate, and provide less resolution power than GC-based methods. Standard deviations in the determination of mass isotopomer distributions with LC-MS-MS are in the range of 6–24% (123), considerably higher than for measurements based on GC-MS analysis of up to 0.4% with, for example, amino acids (124). An improvement in analysis of mass isotopomers from phosphorylated compounds by LC-MS-MS was recently achieved by operating the triple quad in multiple reaction monitoring (MRM) mode (125). Fragmentation of the phosphorylated compounds resulted in high yields of [PO3]$^-$ and/or [H2PO4]$^-$ ions that were subsequently used in deriving the carbon labeling patterns of their parent molecules. MS also usually does not provide positional labeling information. To overcome this limitation, sustained off-resonance irradiation collision-induced dissociation (SORI-CID) has been applied in direct infusion Fourier transform–ion cyclotron resonance mass spectrometry (FT-ICR MS) to fragment the molecules. This way positional information on the ^{13}C label in the proteinogenic amino acids glutamic and aspartic acids was retrieved (126). Another remarkable feature of FT-ICR MS is its high resolution, enabling high-throughput profiling of metabolites and their mass isotopomers without prior chromatographic separation. However, molecules with the same mass (isomers), for example, the amino acids leucine and isoleucine, cannot be distinguished by this method.

NMR spectroscopy can provide detailed information about the structure, dynamics, reaction state, and chemical environment of molecules (127). NMR spectroscopy exploits the physical phenomenon that magnetic nuclei in a magnetic field absorb and re-emit electromagnetic radiation. This energy is at a specific resonance frequency that depends on the strength of the magnetic field and the magnetic properties of the atoms. NMR spectroscopy has been widely applied to elucidating biosynthetic pathways (128,129). ^{13}C labeling patterns can be detected either directly or indirectly through the attached protons. The choice of detecting ^{13}C atoms either directly or indirectly is often determined by balancing increased sensitivity of ^{1}H NMR detection against increased spectral resolution of ^{13}C NMR. Also, ^{1}H NMR allows detection of protons attached to ^{12}C (i.e., total metabolic pools), whereas ^{13}C NMR can assess ^{13}C–^{13}C isotopomer patterns (130).

When using ^{1}H NMR spectroscopy to assess labeling information, ^{13}C enrichments can be calculated from the ratio of the satellite/center peak area of each proton. In a nondecoupled ^{1}H spectrum, signals of ^{13}C-bound protons appear symmetrically as satellite signals around the signal of ^{12}C-bound protons. In the ^{13}C-decoupled ^{1}H spectrum, signals of ^{13}C protons appear at the same position as signals of ^{12}C-bound protons and not as satellite peaks. Therefore, subtracting the ^{13}C-decoupled spectrum from the nondecoupled spectrum allows for accurate quantification of ^{13}C satellite signals areas even with baseline interferences or background signals from ^{12}C-bound protons (131). However, only proton-bound carbons can be investigated. To address this shortcoming, a method for determining ^{13}C enrichments in nonprotonated carbon atoms was developed that makes use of unresolved ^{13}C satellites of proton(s) bonded to the vicinal carbon atom (132).

Low sensitivity, due to low natural abundance of the ^{13}C isotope (1.1%) and a gyromagnetic ratio of only 1/4 that of ^{1}H, had originally restricted ^{13}C NMR analysis. However, the advent of signal-averaging and Fourier transform techniques brought about a dramatic change in the utility of ^{13}C NMR (133). The ^{13}C nucleus exhibits a wide range of chemical shifts, and these shifts are extremely sensitive to the chemical environment. Also, because the gyromagnetic ratio of ^{13}C is small, the relaxation rates of this nucleus are relatively low. The combination of large chemical shifts and favorable relaxation effects, which result in widely shifted groups of narrow ^{13}C resonance, allow for high-resolution ^{13}C NMR experiments in aqueous solutions (134). In contrast to ^{1}H NMR, the intensities of ^{13}C NMR signals are not proportional to the number of equivalent ^{13}C atoms but instead are strongly dependent on the number of surrounding spins (127). Hence for determining ^{13}C enrichments by ^{13}C NMR, first the signal areas of a carbon atom i in an analyzed compound n are determined: $A1_{n,i}$. Subsequently a known amount of the said compound with a known fractional enrichment of ^{13}C in the investigated carbon position, typically a standard solution labeled at natural abundance, is added to the sample solution: $N_{std\ n,i}$. Another ^{13}C spectrum is acquired, and again the (now increased) signal area of the respective carbon is determined: $A2_{n,i}$. With knowledge of the amount of the compound n in the sample from a prior measurement, $N_{meas\ n}$, for example, by high-performance liquid chromatography (HPLC) or GC, the fractional enrichment of the carbon i in compound n can now be calculated from the relative signals areas determined according to $FE = A1_{n,i}/(A2_{n,I} - A1_{n,i}) \times N_{std\ n,i}/N_{meas\ n}$ (131,135). This way Walker et al. obtained the absolute ^{13}C enrichments from ^{1}H NMR spectra and the multiplet intensities from the ^{13}C NMR spectra and used them to conclude that *Microbacterium ammoiziaphilum* synthesizes glutamate mainly via the EMP pathway and the action of phosphoenolpyruvate carboxylase (EC 4.1.1.31) (136). Sonntag et al. (131) used ^{13}C and ^{1}H NMR to determine the relative succinylase flux versus the diaminopimelate dehydrogenase branch in a *C. glutamicum* strain. ^{13}C NMR with proton decoupling was also used to analyze fractional enrichment in phenylalanine. FE data were used to carry out metabolic flux analysis in a phenylalanine-producing *E. coli*. Flux results were compared with optimum flux distributions derived from stoichiometric NA, and overexpression of phosphoenolpyruvate synthetase (EC 2.7.9.2) was identified as a promising metabolic engineering strategy to increase production (137).

Among the large number of heteronuclear NMR schemes, 2D [$^{13}C,^{1}H$]-correlation spectroscopy ([$^{13}C,^{1}H$]-COSY, also referred to as heteronuclear single quantum coherence [HSQC] spectroscopy) uses large one-bond scalar coupling to link carbon chemical shifts with the resonances of directly attached protons. In a 2D [$^{13}C,^{1}H$]-COSY spectrum, the ^{13}C resonance fine structure observed along ω_1 results from the superposition of the fine structures of the isotopomers of the respective metabolite, weighted by their relative abundance. Since 2D [$^{13}C,^{1}H$]-COSY suffices to resolve all relevant resonances, compound mixtures, for example, of amino acids, can be analyzed without prior separation (Figure 2.8). As only one-bond scalar coupling constants, $^{1}J_{CC}$, are large enough to be resolved in the ^{13}C dimension (138), the ^{13}C fine structure of an atom is solely determined by the ^{13}C-labeling pattern of its directly attached neighbor carbon atoms. However, in proton-detected 2D [$^{13}C,^{1}H$]-COSY, only proton-bound carbons can be investigated (129). Emmerling et al. (139) used proton-detected 2D [$^{13}C,^{1}H$]-COSY to show that *E. coli* responds to disruption of both pyruvate kinase isoenzymes by local rerouting of flux via the combined reactions of phosphoenolpyruvate carboxylase and malic enzyme. Proton-detected 2D [$^{13}C,^{1}H$]-COSY was also used to analyze metabolic fluxes of riboflavin producing *B. subtilis* under carbon limited (140) and carbon excess conditions (141), as well as during substrate co-metabolism

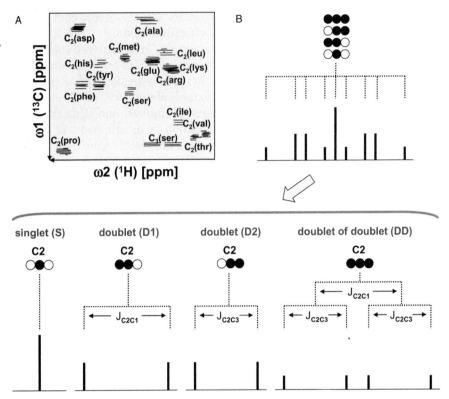

FIGURE 2.8. (A) Region of a 2D $[^{13}C,^{1}H]$-COSY spectrum with the cross peaks of the C2 atoms ($^{13}C2$-$^{1}H2$) of multiple amino acids as well as the C3 peak ($^{13}C3$-$^{1}H2$) of serine according to Reference (186) (B) Idealized cross peak signal of a three-carbon molecule comprising an isomolar mixture of four isotopomers, labeled at either C2, C1 and C2, C2, and C3, or at all carbon positions, respectively. ^{13}C-labeled nuclei are represented by filled circles. The C2 carbon is the observed nucleus. The multiplet pattern is a weighted superposition of singlet (S), doublet (D1 and D2), and doublet of doublet (DD) signals. In case of equal coupling constants $J_{C2C1} = J_{C2C3}$, doublets D1 and D2 cannot be distinguished and the completely labeled C3 fragment will give rise to a triplet instead of a doublet of doublets. Terminal carbons only give singlets or doublets.

(142). Finally, a 2D NMR method based on zero-quantum filtered (ZQF-) TOtal Correlation SpectroscopY (TOCSY) was developed to measure ^{13}C enrichments in complex mixtures of ^{13}C-labeled metabolites. Frequently the ^{1}H NMR spectrum may be too overlapped to obtain a direct measure of ^{13}C enrichment. Using ZQF-TOCSY, more than 30 ^{13}C enrichments were measured in labeled biomass hydrolyzate of *E. coli* without need for prior separation of the metabolites (143).

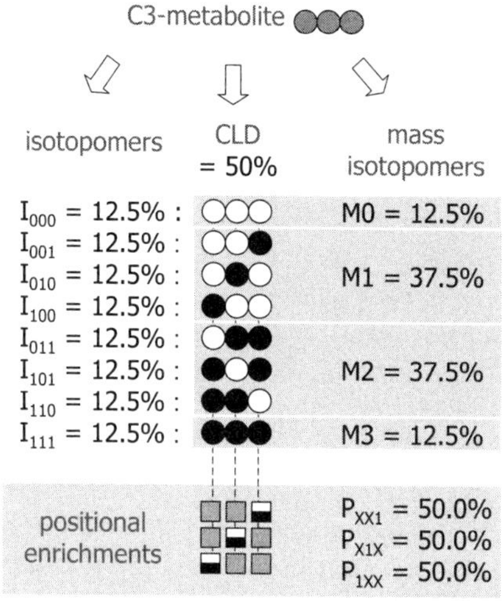

FIGURE 2.9. Four concepts of describing isotope labeling information as exemplified on a three-carbon molecule: isotopomers, CLD (carbon labeling degree), mass isotopomers as well as positional enrichments. Black circles indicate a ^{13}C, white circles, a ^{12}C carbon atom. Boxes indicate positional enrichments, gray boxes or circles indicating positions that are not relevant.

2.7 TRACER-BASED MFA

There are several ways to describe and balance isotope labeling information (Figure 2.9). A basic method is to determine the "carbon labeling degree" (CLD). CLD measurements determine the percentage of an isotope relative to the total amount of atoms of the element in a molecule or molecule fragment. The CLD concept is similar to the determination of "summed fractional enrichment" (144) but requires additional normalization to the number of carbon atoms of the molecule or fragment. The information corresponds to measurements from experiments with radioactive isotopes, for example, with ^{14}C, where the CLD of molecules is determined via scintillation counters or radiograms (145,146). The CLD provides quick information on whether a molecule or fragment contains a label. Molecule- or fragment-based analytical methods such as MS or scintillation counting are best suited to assess CLD. The ^{13}C CLD of naturally occurring carbon molecules or fragments is about 1.1%, equivalent to the natural occurrence of the ^{13}C isotope. Using radioactive isotopes is particularly advantageous if a fraction of the labeled substrate is incorporated in macromolecules and/or cells of which the label content of their various building blocks cannot be assessed and quantified comprehensively. For example, a considerable fraction of glucose is not metabolized by *S. cerevisiae* through either EMP or oxidative PPP, but rather directly incorporated into storage and cell wall polymers. A total ^{14}C balance was successfully applied to quantify flux into these unspecified polymerization reactions (147).

With knowledge on the labeling pattern(s) of a specific metabolite or set of metabolites as well as on the network structure, conclusions on the split ratio of fluxes at a flux branch point can be drawn. Frequently NMR or the combined analysis of several fragments of the same metabolite measured by MS analysis can provide position-specific labeling information, usually termed fractional or positional enrichment (Figure 2.9). In case of a single-carbon molecule such as CO_2, information on the labeling patterns is equivalent to knowing its CLD. With knowledge of topology of the biochemical system and the positional fate of atoms in the enzymatic conversions, measured isotope signals are related by explicit formulas to yield the desired flux ratios. An early application of this "split ratio analysis" was the use of $^{14}C_1$ and $^{14}C_6$ glucose to prove the operation of the oxidative PPP in S. cerevisiae (148). These experiments were based on the fact that if glucose labeled in the C1 position enters the oxidative PPP, labeled CO_2 will be released in the 6-phosphogluconate dehydrogenase (EC 1.1.1.44) reaction. However, CO_2 formed from $^{14}C_6$-labeled glucose in the oxidative PPP will not contain any label. Therefore, the difference in radioactivity in CO_2 formed from $^{14}C_1$ and $^{14}C_6$ provides a measure of the relative flux through the oxidative PPP. Nevertheless, interpretation of the CO_2 labeling pattern also needs to consider substrate recycling, scrambling of label in reversible reactions, and generation of CO_2 from other pathways. There are also mere implementation effects, for example, the buffer effect of fermentation broth or carbon fixation in carboxylating reactions (149). However, while use of radioactive $^{14}CO_2$ is experimentally challenging, good progress was made recently in using $^{13}CO_2$ for flux calculations instead (150). The value of the approach was demonstrated by analyzing fluxes in a lysine-producing C. glutamicum (151).

If not CO_2 but the positional labeling of an intermediate of the triosephosphate pool is analyzed, in addition conclusions on the relative fluxes, not only through the oxidative PPP but also the EDP versus the EMP, can be drawn (Figure 2.10). Sonntag et al. (131) quantified the flux partitioning in the split pathway of lysine synthesis in C. glutamicum by using ^{13}C-NMR spectroscopy to analyze labeling patterns in lysine and in pyruvate-derived metabolites. Another example of the successful use of fractional enrichment measurements was described by Ishino et al. (152). The authors used NMR and the ^{13}C label quantified in the C6 position of histidine to derive an EMP:ox. PPP split ratio of 56:44 in a histidine-producing C. glutamicum strain. This conclusion was possible because the major source of C_1 for histidine formation was derived from serine through action of the serine hydroxymethyltransferase (EC 2.1.2.1). Additional examples for using positional labeling information for flux calculations include the determination of fluxes through the TCA and the glyoxylate shunt in E. coli based on the ^{13}C NMR spectrum of intracellular glutamate (153). Rollin et al. (154) derived multiple flux ratios for the determination of EMP:ox. PPP (55%:45%), anaplerotic pathways (61%), and

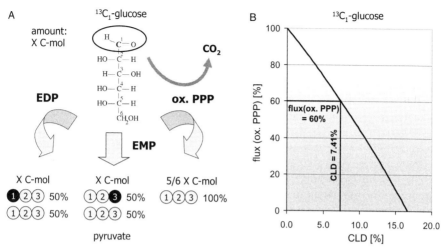

FIGURE 2.10. Split ratio analysis as exemplified by the metabolism of 100% $^{13}C_1$- but otherwise unlabeled glucose either via EDP, EMP, or ox. PPP to pyruvate. Both approaches provide only an approximation, as, for example, pyruvate can also originate from different sources than glucose, as well as due to recycling and exchange reactions in EMP and nonoxidative branch of the PPP that lead to a scrambling of label. (A) If pyruvate is labeled in the C1 position, the molecule was generated from $^{13}C_1$-glucose via the EDP. If pyruvate is labeled in the C3 position, the molecule was generated from $^{13}C_1$-glucose via the EMP. As both pathways split the six-carbon glucose molecule into two pyruvate, but only one pyruvate carries the label, the other 50% of pyruvate molecules resulting from these pathways will not be labeled. If $^{13}C_1$-glucose is metabolized via the ox. PPP, the $^{13}C_1$ atom of glucose is released as CO_2 and the resulting five-carbon molecules are subsequently rearranged to yield pyruvate molecules without any isotope label. However, as 1/6 of the carbon originally provided in glucose is lost to CO_2, the total amount of unlabeled pyruvate produced will be only containing 5/6 of the original amount of carbon in glucose. Consequently if the isotopomer composition of pyruvate can be determined, conclusions on the activity of the three pathways can be drawn. (B) If only the CLD of pyruvate can be measured, the split ratio of the ox. PPP can be directly derived, as exemplified with a measured CLD of 7.4%, which corresponds to a molar flux through the ox. PPP of 60%. However, no conclusions on the relative activities of the EDP and EMP can be drawn.

glyoxylate shunt (0%) from fractional enrichment measurements in glutamate in *Corynebacterium melassecola*.

Sauer et al. (155) finally expanded the use of explicit equations to cover several important flux ratios in central carbon metabolism and termed the approach metabolic flux ratio analysis (METAFoR). Flux ratios were derived from either extensive NMR or MS analyses (156,157). The approach is computationally inexpensive and can be applied for high-throughput analysis of important flux nodes in metabolic networks (155). Flux ratios can also be used as additional constraints in stoichiometric MFA (158–160). However, a major drawback of the approach is that only a small fraction of the information gained from analyzing labeling patterns is exploited. Moreover, no comprehensive statistical method can be deployed to assess the quality of the calculated flux distributions. This capability is especially helpful if the network topology of the biochemical system is not known so that assumptions made for deriving the analytical equations may not hold valid. Rantanen et al. (161) developed a computational framework to derive METAFoR constraints automatically and comprehensively for any combination of substrates and isotope measurements. Previously, deriving constraints required knowledge of domain experts.

Zupke and Stephanopoulos (162) introduced atom mapping matrices, a clear and intuitive mathematical formalism that allowed balancing of fractional enrichments based on knowledge of the biochemical system and the fate of atoms in the enzymatic reactions. Marx et al. (163) carried out metabolic flux analysis in a lysine-producing *C. glutamicum* strain in continuous culture at a dilution rate of $D = 0.1$ 1/h, based on a comprehensive model comprising balances on metabolites and fractional enrichment data. Fractional enrichments in 11 amino acids at 31 positions were determined. For this purpose amino acids were purified from hydrolyzates of cellular protein by cation exchange chromatography and analyzed by ^{1}H NMR. With the knowledge of their biosynthesis pathway, the positional enrichment in the amino acids allowed the authors to draw conclusions on the fractional enrichment of 20 positions of six metabolites in central carbon metabolism. In particular, all carbon positions of erythrose-4-phosphate, glyceraldehyde-3-phosphate, pyruvate, alpha-ketoglutarate, oxaloacetate, and carbon dioxide were resolved. Moreover, fractional enrichments in all positions in lysine were determined. A high pentose phosphate flux of 66.4% of the molar glucose uptake rate, an anaplerotic pyruvate carboxylase (EC 6.4.1.1) flux of 38%, and a relative flux of the succinylase as compared with the diaminopimelate dehydrogenase branch of 2.8:1.0 in the lysine biosynthesis pathway were identified, respectively. High exchange fluxes in the nonoxidative branch of the PPP were discovered, in particular in the 5-phosphate transketolase (EC 2.2.1.1) reaction. In another study, Christensen and Nielsen (144) grew *Penicillium chrysogenum* on a defined medium with ${}^{13}C_1$-glucose as the sole carbon and energy source, and added phenoxyacetic acid as side-chain precursor for the biosynthesis of penicillin V. By balancing fractional enrichments and metabolites they found that glycine was synthesized not only by serine hydroxymethyltransferase, but also by threonine aldolase. The authors also detected that acetyl-coenzyme A (acetyl-CoA) was derived not only from citrate via the ATP citrate lyase reaction (EC 2.3.3.8), but also from the degradation of the penicillin side-chain precursor, phenoxyacetic acid. Finally, Christensen et al. also balanced fractional enrichment and metabolite measurements to carry out a flux analysis in an aerobic chemostat culture of *S. cerevisiae* at $D = 0.1$ 1/h and found an EMP:ox PPP flux ratio of 35:43 and an anaplerotic reaction via carboxylation of pyruvate of about 26% of the specific glucose uptake rate (164). In both studies the authors balanced fractional enrichments. However, as GC-MS

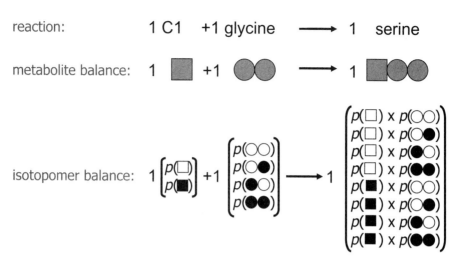

FIGURE 2.11. Isotopomer balancing exemplified on hand of the reaction of a one- and two- to three-carbon atom molecule. For balancing the isotopomers, fractions (probabilities p) of isotopomers in the one-carbon molecule are multiplied with the fractions (probabilities p) of the isotopomer of the two-carbon atoms to yield the fractions (probabilities p) of the respective isotopomers of the three-carbon atom. Black squares and circles illustrate the ^{13}C, white squares and circles, the ^{12}C isotope.

analysis was applied to assess mass isotopomer fractions of metabolites, summed fractional enrichments were used to judge the quality of the fitted results.

Maximum information retrieval from NMR and MS data obtained in a tracer experiment is accomplished by using "isotopomer" (140,165–167) or "cumomer" balances (168). In isotopomer balancing, not the fraction of a labeled carbon atom in a given position of a metabolite is balanced, but the fraction of isotopomers. Consequently in a bimolecular reaction of molecules A and B reacting to C, the fraction of a specific isotopomer in C results from the summation of the products of every single isotopomer fraction in A multiplied with every single isotopomer fraction in B that yields the said specific isotopomer of C in the reaction (Figure 2.11). If several reactions can form molecule C, the isotopomer probability of each reaction need now to be multiplied with the magnitude of the flux through the respective reaction, added up, and subsequently normalized to the sum of all fluxes generating the metabolite. The mathematical formulation of isotopomer balances therefore results in large and nonlinear terms. The basic principles behind these approaches as well as their advantages and limitations have been reviewed by Wiechert (169). An example of MFA based on isotopomer balances is given in Figure 2.12.

Antoniewicz et al. (170) introduced the "elementary metabolite units" or EMU framework for balancing isotope labels in metabolic flux networks. The essence of the EMU approach is that only the labeling information that is required to describe obtained measurement data is balanced throughout the

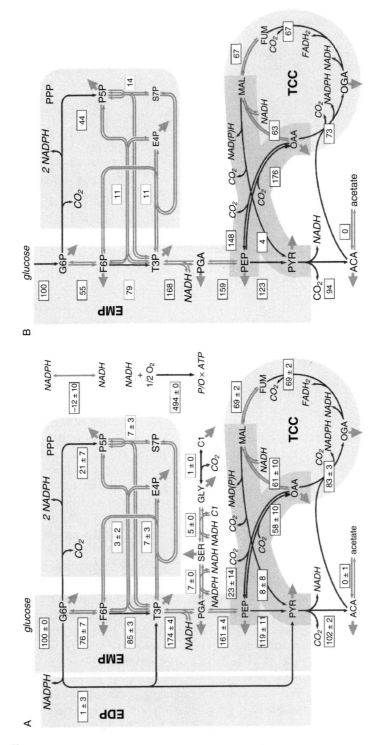

FIGURE 2.12. Intracellular flux distributions in *E. coli* as derived from comprehensive isotopomer balancing models. (A) Chemostat culture in steady state at $D = 0.22$ L/h. Labeling data were generated by GC-MS and COSY NMR and used in metabolic and isotopic stationary MFA (165). (B) Chemostat culture in steady state at $D = 0.10$ L/h. Labeling data were generated by GC-MS and LC-MS and used in metabolic stationary but isotopic instationary MFA (95).

network. This way the approach allows for a significant reduction of balance equations as compared with isotopomer or cumomer frameworks. For a typical ^{13}C labeling system, the total number of variables and equations that needs to be solved can be reduced by one order of magnitude. The significantly reduced number of balance equations now also makes the simultaneous description of multiple elemental isotopes in a molecule possible, for example, in glucose the combination of $^{1}H/^{2}H$, $^{12}C/^{13}C$, and $^{16}O/^{18}O$. In this case the analysis of gluconeogenesis, for example, requires only 354 EMU balances, compared with more than 2 million isotopomer balances.

Tracer-based MFA under isotopic or metabolic instationary conditions (Figure 2.5) is still in the realm of method development and will require further refinement before it is broadly applied in industrial R&D (171). This is due to the dramatic effort necessary to generate and analyze valuable samples as well as the complexity of extracting information by modeling and simulation. Nevertheless, analysis of the isotopomer time profiles of metabolites under "isotopically instationary" conditions enables the determination of fluxes with improved accuracy, and can also be used in predicting intracellular concentrations of metabolites that cannot be measured with the available analytical procedures (93–96). Noh and colleagues (122) applied isotopically nonstationary flux analysis in a fed-batch cultivation of *E. coli*. Solving the differential equation system was computationally expensive, with one simulation requiring 3–8 minutes. A 5000-fold reduction in simulation time, from 83 minutes with cumomer balancing to 1 second, was achieved by adapting the EMU framework to nonstationary flux analysis and applying "block decoupling" to decompose the EMU system into smaller sub-problems (172). Estimating fluxes and concentrations from 33 GC-MS fragments in a large *E. coli* metabolic network of 35 free fluxes and 46 metabolite pools took only 15 minutes, opening a broad range of new applications for SREs and isotopically nonstationary flux analysis. Even more information can be obtained in SRE experiments with isotopic tracers (105) (Figure 2.5).

2.8 VALIDATING METABOLIC FLUX NETWORKS

Conventionally assumptions on the topology of a metabolic flux network are confirmed by *in vitro* enzymatic assays or phenotypic comparison of gene deletion strains. In industrial biotechnology, typically the validation of networks are required on two occasions: (i) to make sure that the basic assumptions on the metabolic flux network is correct before engaging in an extensive metabolic engineering project, and (ii) to trace the effect of a genetic modification on a metabolic flux system. Because if the gene deletion is just a first step in a multi-step genetic engineering strategy and either (i) does not cause a change in phenotype, (ii) no assay for the enzyme of interest exists, or (iii) the enzyme would have to be purified first (e.g., because multiple enzymes in the crude extract consume the assay substrates with high rate), tracer-based analysis may provide the fastest and most accurate way for hypothesis testing. The deletion of the two pyruvate kinase (EC 2.7.1.40) enzymes in *E. coli*, for example, did not result in any discernible phenotype. However, analysis of the intracellular fluxes by ^{13}C MFA with a comprehensive isotopomer balancing model did reveal the gene deletion. It was shown that the blocked pyruvate kinase flux was rerouted via the combined reactions of phosphoenolpyruvate carboxylase (EC 4.1.1.31), malate dehydrogenase (EC 1.1.1.37) and malic enzyme (EC 1.1.1.39) (139). Also in *E. coli*, the transaldolase (EC 2.2.1.2) gene was deleted and the mutant was subsequently grown on xylose. However, in contrast to conventional textbook knowledge, a good fit of the labeling data was only accomplished by the introduction of a new reaction sequence (37). It was suggested that phosphofructokinase (EC 2.7.1.11) and fructose-bisphosphate aldolase (EC 4.1.2.13) catalyzed the conversion of sedoheptulose-7-phosphate to sedoheptulose-1,7-bisphosphate and subsequently to erythrose-4-phosphate and dihydroxyacetone phosphate, respectively. Further support for the postulated alternative pathway was obtained by metabolite measurements and *in vitro* enzyme assays. In *P. chrysogenum* and *Aspergillus nidulans* ^{13}C-based MFA revealed labeling patterns in cytosolic acetyl-CoA that could not be explained by operation of EMP and PPP only (173). In a subsequent study, an active phosphoketolase pathway was detected (174). More examples of the use of ^{13}C MFA for the exploration of metabolic flux networks can be found in References (175) and (176).

Constraints derived from thermodynamic analyses provide a way of checking consistency of the obtained flux solutions (177–179). In order for a reaction (or flux) to occur, the change of the associated Gibb's free reaction energy ΔG needs to be negative. Gibb's free reaction energy depends on the chemical properties of the reaction partners—their standard Gibbs energy of formation ΔG_f°—as well as on their activities. Consequently if the chemical properties and activities are known, conclusions on the feasibility of a reaction can be drawn (66,177). The same principles govern exchange fluxes quantified in isotope labeling experiments (180,181). However, activities do not depend only on the measured intracellular metabolite concentrations, temperature, and pressure, but also on other components in the mixture, and consequently their determination represents a major source of uncertainty. Moreover, many compounds occurring in biological systems are not available as pure substrates, and consequently their ΔG_f° values cannot be determined experimentally. Methods to estimate ΔG_f° values based on group allocation theories were developed (182,183). Correcting functions as well as improved methods to estimate ΔG_f° were introduced to improve the accuracy of ΔG determinations (184,185). Nonetheless, a wide range of feasible metabolite concentrations for a given flux phenotype (66) indicates that vice versa the assessment of the feasibility of a flux based on error-prone intracellular metabolite concentrations is only of limited value and should be treated with care. More details on network embedded thermodynamic (NET) analysis and thermodynamics-based metabolic flux (TMFA) analysis are given in Chapter 3.

2.9 CONCLUSIONS

Analysis and design of metabolic flux networks is at the core of industrial biotechnology, as fluxes result in yields and titers, and it is primarily rate, titer, and yield that contribute to the commercial success of novel products or processes. Metabolic engineering provides powerful tools for the modification, analysis, and design of complex metabolic systems. The advent of "omics" technologies and the integration of tracer experiments into NA and MFA considerably expanded our understanding of systems and allows delivery of fast and accurate predictions. The availability of fully representative *in silico* cell models is merely a matter of time and will result in a swift and fundamental paradigm change in metabolic engineering.

ACKNOWLEDGMENTS

Critical review of the chapter by Chris Roe and Brian Lefebvre is gratefully acknowledged.

REFERENCES

1. Dauner, M. (2010) From fluxes and isotope labeling patterns towards *in silico* cells. *Curr Opin Biotechnol*, **21**, 55–62.
2. Hutchison, C.A., 3rd (2007) DNA sequencing: bench to bedside and beyond. *Nucleic Acids Res*, **35**, 6227–6237.
3. Falk, R., Ramstroem, M., Stahl, S., and Hober, S. (2007) Approaches for systematic proteome exploration. *Biomol Eng*, **24**, 155–168.
4. van der Werf, M.J., Jellema, R.H., and Hankemeier, T. (2005) Microbial metabolomics: replacing trial-and-error by the unbiased selection and ranking of targets. *J Ind Microbiol Biotechnol*, **32**, 234–252.
5. Schena, M., Shalon, D., Davis, R.W., and Brown, P.O. (1995) Quantitative monitoring of gene expression patterns with a complementary DNA microarray. *Science*, **270**, 467–470.
6. Metzker, M.L. (2010) Sequencing technologies—the next generation. *Nat Rev Genet*, **11**, 31–46.
7. Fleischmann, R.D., Adams, M.D., White, O., Clayton, R.A., Kirkness, E.F., Kerlavage, A.R., Bult, C.J., Tomb, J.F., Dougherty, B.A., Merrick, J.M. et al. (1995) Whole-genome random sequencing and assembly of *Haemophilus influenzae* Rd. *Science*, **269**, 496–512.
8. Venter, J.C., Adams, M.D., Myers, E.W., Li, P.W., Mural, R.J., Sutton, G.G., Smith, H.O., Yandell, M., Evans, C.A., Holt, R.A. et al. (2001) The sequence of the human genome. *Science*, **291**, 1304–1351.
9. Lander, E.S., Linton, L.M., Birren, B., Nusbaum, C., Zody, M.C., Baldwin, J., Devon, K., Dewar, K., Doyle, M., FitzHugh, W. et al. (2001) Initial sequencing and analysis of the human genome. *Nature*, **409**, 860–921.

10. DeRisi, J.L., Iyer, V.R., and Brown, P.O. (1997) Exploring the metabolic and genetic control of gene expression on a genomic scale. *Science*, **278**, 680–686.

11. Lamesch, P., Dreher, K., Swarbreck, D., Sasidharan, R., Reiser, L., and Huala, E. (2010) Using the Arabidopsis information resource (TAIR) to find information about Arabidopsis genes. *Curr Protoc Bioinformatics*, **Chapter 1**, Unit1 11.

12. VanBogelen, R.A., Abshire, K.Z., Moldover, B., Olson, E.R., and Neidhardt, F.C. (1997) *Escherichia coli* proteome analysis using the gene-protein database. *Electrophoresis*, **18**, 1243–1251.

13. Hecker, M. (2003) A proteomic view of cell physiology of *Bacillus subtilis*—bringing the genome sequence to life. *Adv Biochem Eng Biotechnol*, **83**, 57–92.

14. Maruyama, Y., Kawamura, Y., Nishikawa, T., Isogai, T., Nomura, N., and Goshima, N. (2012) HGPD: human gene and protein database, 2012 update. *Nucleic Acids Res*, **40**, D924–D929.

15. Rabinowitz, J.D. (2007) Cellular metabolomics of *Escherichia coli*. *Expert Rev Proteomics*, **4**, 187–198.

16. Fiehn, O., Kopka, J., Dormann, P., Altmann, T., Trethewey, R.N., and Willmitzer, L. (2000) Metabolite profiling for plant functional genomics. *Nat Biotechnol*, **18**, 1157–1161.

17. Wishart, D.S., Knox, C., Guo, A.C., Eisner, R., Young, N., Gautam, B., Hau, D.D., Psychogios, N., Dong, E., Bouatra, S. et al. (2009) HMDB: a knowledge base for the human metabolome. *Nucleic Acids Res*, **37**, D603–D610.

18. Strelkov, S., von Elstermann, M., and Schomburg, D. (2004) Comprehensive analysis of metabolites in *Corynebacterium glutamicum* by gas chromatography/mass spectrometry. *Biol Chem*, **385**, 853–861.

19. Kaitin, K.I. (2010) Deconstructing the drug development process: the new face of innovation. *Clin Pharmacol Ther*, **87**, 356–361.

20. van der Greef, J. and McBurney, R.N. (2005) Rescuing drug discovery: *in vivo* systems pathology and systems pharmacology. *Nat Rev Drug Discov*, **4**, 961–967.

21. Hubner, K., Sahle, S., and Kummer, U. (2011) Applications and trends in systems biology in biochemistry. *FEBS J*, **278**, 2767–2857.

22. Bailey, J.E. (1991) Toward a science of metabolic engineering. *Science*, **252**, 1668–1675.

23. Lee, J.W., Kim, T.Y., Jang, Y.S., Choi, S., and Lee, S.Y. (2011) Systems metabolic engineering for chemicals and materials. *Trends Biotechnol*, **29**, 370–378.

24. Na, D., Kim, T.Y., and Lee, S.Y. (2010) Construction and optimization of synthetic pathways in metabolic engineering. *Curr Opin Microbiol*, **13**, 363–370.

25. Hibbert, E.G., Baganz, F., Hailes, H.C., Ward, J.M., Lye, G.J., Woodley, J.M., and Dalby, P.A. (2005) Directed evolution of biocatalytic processes. *Biomol Eng*, **22**, 11–19.

26. Tyo, K.E., Kocharin, K., and Nielsen, J. (2010) Toward design-based engineering of industrial microbes. *Curr Opin Microbiol*, **13**, 255–262.

27. Feist, A.M., Herrgard, M.J., Thiele, I., Reed, J.L., and Palsson, B.O. (2009) Reconstruction of biochemical networks in microorganisms. *Nat Rev Microbiol*, **7**, 129–143.

28. Ruder, K. and Winstead, E.R. (2000) Genome News Network.

29. Li, Q.R., Carvunis, A.R., Yu, H., Han, J.D., Zhong, Q., Simonis, N., Tam, S., Hao, T., Klitgord, N.J., Dupuy, D. et al. (2008) Revisiting the *Saccharomyces cerevisiae* predicted ORFeome. *Genome Res*, **18**, 1294–1303.

30. Christie, K.R., Hong, E.L., and Cherry, J.M. (2009) Functional annotations for the *Saccharomyces cerevisiae* genome: the knowns and the known unknowns. *Trends Microbiol*, **17**, 286–294.

31. Brodie, A.F. and Lipmann, F. (1955) Identification of a gluconolactonase. *J Biol Chem*, **212**, 677–685.

32. Park, H.S., Xing, R., and Whitman, W.B. (1995) Nonenzymatic acetolactate oxidation to diacetyl by flavin, nicotinamide and quinone coenzymes. *Biochim Biophys Acta*, **1245**, 366–370.

33. Ozturk, S.S. and Palsson, B.O. (1990) Chemical decomposition of glutamine in cell culture media: effect of media type, pH, and serum concentration. *Biotechnol Prog*, **6**, 121–128.

34. Nagase, S., Takemura, K., Ueda, A., Hirayama, A., Aoyagi, K., Kondoh, M., and Koyama, A. (1997) A novel nonenzymatic pathway for the generation of nitric oxide by the reaction of hydrogen peroxide and D- or L-arginine. *Biochem Biophys Res Commun*, **233**, 150–153.

35. Blackburn, M.L., Ketterer, B., Meyer, D.J., Juett, A.M., and Bull, A.W. (1997) Characterization of the enzymatic and nonenzymatic reaction of 13-oxooctadecadienoic acid with glutathione. *Chem Res Toxicol*, **10**, 1364–1371.

36. Hult, K. and Berglund, P. (2007) Enzyme promiscuity: mechanism and applications. *Trends Biotechnol*, **25**, 231–238.

37. Nakahigashi, K., Toya, Y., Ishii, N., Soga, T., Hasegawa, M., Watanabe, H., Takai, Y., Honma, M., Mori, H., and Tomita, M. (2009) Systematic phenome analysis of *Escherichia coli* multiple-knockout mutants reveals hidden reactions in central carbon metabolism. *Mol Syst Biol*, **5**, 306.

38. Fillingame, R.H. (1997) Coupling H+ transport and ATP synthesis in F1F0-ATP synthases: glimpses of interacting parts in a dynamic molecular machine. *J Exp Biol*, **200**, 217–224.

39. Cross, R.L. and Muller, V. (2004) The evolution of A-, F-, and V-type ATP synthases and ATPases: reversals in function and changes in the H+/ATP coupling ratio. *FEBS Lett*, **576**, 1–4.

40. Saraste, M. (1999) Oxidative phosphorylation at the fin de siecle. *Science*, **283**, 1488–1493.

41. Kadenbach, B. (2003) Intrinsic and extrinsic uncoupling of oxidative phosphorylation. *Biochim Biophys Acta*, **1604**, 77–94.

42. Jarmuszkiewicz, W., Woyda-Ploszczyca, A., Antos-Krzeminska, N., and Sluse, F.E. (2010) Mitochondrial uncoupling proteins in unicellular eukaryotes. *Biochim Biophys Acta*, **1797**, 792–799.

43. Jastroch, M., Divakaruni, A.S., Mookerjee, S., Treberg, J.R., and Brand, M.D. (2010) Mitochondrial proton and electron leaks. *Essays Biochem*, **47**, 53–67.

44. Haines, T.H. (2001) Do sterols reduce proton and sodium leaks through lipid bilayers? *Prog Lipid Res*, **40**, 299–324.

45. Fuhrer, T. and Sauer, U. (2009) Different biochemical mechanisms ensure network-wide balancing of reducing equivalents in microbial metabolism. *J Bacteriol*, **191**, 2112–2121.

46. van der Werf, M.J., Overkamp, K.M., Muilwijk, B., Coulier, L., and Hankemeier, T. (2007) Microbial metabolomics: toward a platform with full metabolome coverage. *Anal Biochem*, **370**, 17–25.

47. Oldiges, M. and Takors, R. (2005) Applying metabolic profiling techniques for stimulus-response experiments: chances and pitfalls. *Adv Biochem Eng Biotechnol*, **92**, 173–196.

48. Cakir, T., Hendriks, M.M., Westerhuis, J.A., and Smilde, A.K. (2009) Metabolic network discovery through reverse engineering of metabolome data. *Metabolomics*, **5**, 318–329.

49. Hatzimanikatis, V., Li, C., Ionita, J.A., Henry, C.S., Jankowski, M.D., and Broadbelt, L.J. (2005) Exploring the diversity of complex metabolic networks. *Bioinformatics*, **21**, 1603–1609.

50. Ugi, I., Bauer, J., Brandt, J., Friedrich, J., Gasteiger, J., Jochum, C., and Schubert, W. (1979) New applications of computers in chemistry. *Angew Chem*, **18**, 111–123.

51. Henry, C.S., Broadbelt, L.J., and Hatzimanikatis, V. (2010) Discovery and analysis of novel metabolic pathways for the biosynthesis of industrial chemicals: 3-hydroxypropanoate. *Biotechnol Bioeng*, **106**, 462–473.

52. Yim, H., Haselbeck, R., Niu, W., Pujol-Baxley, C., Burgard, A., Boldt, J., Khandurina, J., Trawick, J.D., Osterhout, R.E., Stephen, R. et al. (2011) Metabolic engineering of *Escherichia coli* for direct production of 1,4-butanediol. *Nat Chem Biol*, **7**, 445–452.

53. Cedrone, F., Menez, A., and Quemeneur, E. (2000) Tailoring new enzyme functions by rational redesign. *Curr Opin Struct Biol*, **10**, 405–410.

54. Pantazes, R.J., Grisewood, M.J., and Maranas, C.D. (2011) Recent advances in computational protein design. *Curr Opin Struct Biol*, **21**, 467–472.

55. Rothlisberger, D., Khersonsky, O., Wollacott, A.M., Jiang, L., DeChancie, J., Betker, J., Gallaher, J.L., Althoff, E.A., Zanghellini, A., Dym, O. et al. (2008) Kemp elimination catalysts by computational enzyme design. *Nature*, **453**, 190–195.

56. Siegel, J.B., Zanghellini, A., Lovick, H.M., Kiss, G., Lambert, A.R., St Clair, J.L., Gallaher, J.L., Hilvert, D., Gelb, M.H., Stoddard, B.L. et al. (2010) Computational design of an enzyme catalyst for a stereoselective bimolecular Diels-Alder reaction. *Science*, **329**, 309–313.

57. Wiechert, W. and Noack, S. (2011) Mechanistic pathway modeling for industrial biotechnology: challenging but worthwhile. *Curr Opin Biotechnol*, **22**, 604–610.

58. Papoutsakis, E.T. (1984) Equations and calculations for fermentations of butyric acid bacteria. *Biotechnol Bioeng*, **26**, 174–187.

59. Dauner, M. and Sauer, U. (2001) Stoichiometric growth model for riboflavin-producing *Bacillus subtilis*. *Biotechnol Bioeng*, **76**, 132–143.

60. Nissen, T.L., Schulze, U., Nielsen, J., and Villadsen, J. (1997) Flux distributions in anaerobic, glucose-limited continuous cultures of *Saccharomyces cerevisiae*. *Microbiology*, **143**(Pt 1), 203–218.

61. Merino, M.P., Andrews, B.A., and Asenjo, J.A. (2011) Stoichiometric model and metabolic flux analysis for *Leptospirillum ferrooxidans*. *Biotechnol Bioeng*, **107**, 696–706.

62. Varma, A., Boesch, B.W., and Palsson, B.O. (1993) Biochemical production capabilities of *Escherichia coli*. *Biotechnol Bioeng*, **42**, 59–73.

63. Klinke, S., Dauner, M., Scott, G., Kessler, B., and Witholt, B. (2000) Inactivation of isocitrate lyase leads to increased production of medium-chain-length poly(3-hydroxyalkanoates) in Pseudomonas putida. *Appl Environ Microbiol*, **66**, 909–913.

64. Joyce, A.R., Reed, J.L., White, A., Edwards, R., Osterman, A., Baba, T., Mori, H., Lesely, S.A., Palsson, B.O., and Agarwalla, S. (2006) Experimental and computational assessment of conditionally essential genes in *Escherichia coli*. *J Bacteriol*, **188**, 8259–8271.

65. Satish Kumar, V., Dasika, M.S., and Maranas, C.D. (2007) Optimization based automated curation of metabolic reconstructions. *BMC Bioinformatics*, **8**, 212.

66. Henry, C.S., Jankowski, M.D., Broadbelt, L.J., and Hatzimanikatis, V. (2006) Genome-scale thermodynamic analysis of *Escherichia coli* metabolism. *Biophys J*, **90**, 1453–1461.

67. Jorgensen, H., Nielsen, J., Villadsen, J., and Mollgaard, H. (1995) Metabolic flux distributions in *Penicillium chrysogenum* during fed-batch cultivations. *Biotechnol Bioeng*, **46**, 117–131.

68. Vickers, C.E., Klein-Marcuschamer, D., and Kromer, J.O. (2012) Examining the feasibility of bulk commodity production in *Escherichia coli*. *Biotechnol Lett*, **34**, 585–596.

69. Varman, A.M., Xiao, Y., Leonard, E., and Tang, Y.J. (2011) Statistics-based model for prediction of chemical biosynthesis yield from *Saccharomyces cerevisiae*. *Microb Cell Fact*, **10**, 45.

70. Matsuda, F., Furusawa, C., Kondo, T., Ishii, J., Shimizu, H., and Kondo, A. (2011) Engineering strategy of yeast metabolism for higher alcohol production. *Microb Cell Fact*, **10**, 70.

71. Schuster, S. and Schuster, R. (1991) Detecting strictly detailed balanced subnetworks in open chemical reaction networks. *J Math Chem*, **6**, 17–40.

72. Gayen, K. and Venkatesh, K.V. (2006) Analysis of optimal phenotypic space using elementary modes as applied to *Corynebacterium glutamicum*. *BMC Bioinformatics*, **7**, 445.

73. Carlson, R., Fell, D., and Srienc, F. (2002) Metabolic pathway analysis of a recombinant yeast for rational strain development. *Biotechnol Bioeng*, **79**, 121–134.

74. Chen, Z., Liu, H., Zhang, J., and Liu, D. (2010) Elementary mode analysis for the rational design of efficient succinate conversion from glycerol by *Escherichia coli*. *J Biomed Biotechnol*, **2010**, 518743.

75. Klamt, S. and Stelling, J. (2002) Combinatorial complexity of pathway analysis in metabolic networks. *Mol Biol Rep*, **29**, 233–236.

76. Trinh, C.T., Wlaschin, A., and Srienc, F. (2009) Elementary mode analysis: a useful metabolic pathway analysis tool for characterizing cellular metabolism. *Appl Microbiol Biotechnol*, **81**, 813–826.

77. Varma, A. and Palsson, B.O. (1994) Metabolic flux balancing: basic concepts, scientific and practical use. *Nat Biotech*, **12**, 994–998.

78. Stephanopoulos, G.N., Aristidou, A.A., and Nielsen, J. (1998) *Metabolic Engineering. Principles and Methodologies.* Academic Press, London.

79. Lange, H.C. and Heijnen, J.J. (2001) Statistical reconciliation of the elemental and molecular biomass composition of *Saccharomyces cerevisiae*. *Biotechnol Bioeng*, **75**, 334–344.

80. von Stockar, U., Gustafsson, L., Larsson, C., Marison, I., Tissot, P., and Gnaiger, E. (1993) Thermodynamic considerations in constructing energy balances for cellular growth. *Biochim Biophys Acta Bioenerg*, **1183**, 221–240.

81. van der Heijden, R.T., Heijnen, J.J., Hellinga, C., Romein, B., and Luyben, K.C. (1994) Linear constraint relations in biochemical reaction systems: I. Classification of the calculability and the balanceability of conversion rates. *Biotechnol Bioeng*, **43**, 3–10.

82. Bonarius, H.P.J., Schmid, G., and Tramper, J. (1997) Flux analysis of underdetermined metabolic networks: the quest for the missing constraints. *Trends Biotechnol*, **15**, 308–314.

83. Mollney, M., Wiechert, W., Kownatzki, D., and de Graaf, A.A. (1999) Bidirectional reaction steps in metabolic networks: IV. Optimal design of isotopomer labeling experiments. *Biotechnol Bioeng*, **66**, 86–103.

84. Wiechert, W. and Noh, K. (2005) From stationary to instationary metabolic flux analysis. *Adv Biochem Eng Biotechnol*, **92**, 145–172.

85. Bartek, T., Blombach, B., Lang, S., Eikmanns, B.J., Wiechert, W., Oldiges, M., Noh, K., and Noack, S. (2011) Comparative 13C metabolic flux analysis of pyruvate dehydrogenase complex-deficient, L-valine-producing *Corynebacterium glutamicum*. *Appl Environ Microbiol*, **77**, 6644–6652.

86. Laney, J.D. and Hochstrasser, M. (2004) Ubiquitin-dependent control of development in *Saccharomyces cerevisiae*. *Curr Opin Microbiol*, **7**, 647–654.

87. Grotkjaer, T., Akesson, M., Christensen, B., Gombert, A.K., and Nielsen, J. (2004) Impact of transamination reactions and protein turnover on labeling dynamics in (13)C-labeling experiments. *Biotechnol Bioeng*, **86**, 209–216.

88. Iwatani, S., Van Dien, S., Shimbo, K., Kubota, K., Kageyama, N., Iwahata, D., Miyano, H., Hirayama, K., Usuda, Y., Shimizu, K. et al. (2007) Determination of metabolic flux changes during fed-batch cultivation from measurements of intracellular amino acids by LC-MS/MS. *J Biotechnol*, **128**, 93–111.

89. Shaikh, A.S., Tang, Y.J., Mukhopadhyay, A., and Keasling, J.D. (2008) Isotopomer distributions in amino acids from a highly expressed protein as a proxy for those from total protein. *Anal Chem*, **80**, 886–890.

90. Jonsbu, E., Christensen, B., and Nielsen, J. (2001) Changes of *in vivo* fluxes through central metabolic pathways during the production of nystatin by *Streptomyces noursei* in batch culture. *Appl Microbiol Biotechnol*, **56**, 93–100.

91. Fischer, E., Zamboni, N., and Sauer, U. (2004) High-throughput metabolic flux analysis based on gas chromatography-mass spectrometry derived 13C constraints. *Anal Biochem*, **325**, 308–316.

92. Fischer, E. and Sauer, U. (2005) Large-scale *in vivo* flux analysis shows rigidity and suboptimal performance of *Bacillus subtilis* metabolism. *Nat Genet*, **37**, 636–640.

93. Noh, K., Wahl, A., and Wiechert, W. (2006) Computational tools for isotopically instationary 13C labeling experiments under metabolic steady state conditions. *Metab Eng*, **8**, 554–577.

94. Noh, K. and Wiechert, W. (2006) Experimental design principles for isotopically instationary 13C labeling experiments. *Biotechnol Bioeng*, **94**, 234–251.

95. Schaub, J., Mauch, K., and Reuss, M. (2008) Metabolic flux analysis in *Escherichia coli* by integrating isotopic dynamic and isotopic stationary 13C labeling data. *Biotechnol Bioeng*, **99**, 1170–1185.

96. Zhao, Z., Kuijvenhoven, K., Ras, C., van Gulik, W.M., Heijnen, J.J., Verheijen, P.J., and van Winden, W.A. (2008) Isotopic non-stationary 13C gluconate tracer method for accurate determination of the pentose phosphate pathway split-ratio in *Penicillium chrysogenum*. *Metab Eng*, **10**, 178–186.

97. Aboka, F.O., Heijnen, J.J., and van Winden, W.A. (2009) Dynamic 13C-tracer study of storage carbohydrate pools in aerobic glucose-limited *Saccharomyces cerevisiae* confirms a rapid steady-state turnover and fast mobilization during a modest stepup in the glucose uptake rate. *FEMS Yeast Res*, **9**, 191–201.

98. Antoniewicz, M.R., Kraynie, D.F., Laffend, L.A., Gonzalez-Lergier, J., Kelleher, J.K., and Stephanopoulos, G. (2007) Metabolic flux analysis in a nonstationary system: fed-batch fermentation of a high yielding strain of *E. coli* producing 1,3-propanediol. *Metab Eng*, **9**, 277–292.

99. Ruhl, M., Zamboni, N., and Sauer, U. (2009) Dynamic flux responses in riboflavin overproducing *Bacillus subtilis* to increasing glucose limitation in fed-batch culture. *Biotechnol Bioeng*, **105**, 795–804.

100. Theobald, U., Mailinger, W., Baltes, M., Rizzi, M., and Reuss, M. (1997) *In vivo* analysis of metabolic dynamics in *Saccharomyces cerevisiae*: I. Experimental observations. *Biotechnol Bioeng*, **55**, 305–316.

101. Schaub, J., Schiesling, C., Reuss, M., and Dauner, M. (2006) Integrated sampling procedure for metabolome analysis. *Biotechnol Prog*, **22**, 1434–1442.

102. Schadel, F. and Franco-Lara, E. (2009) Rapid sampling devices for metabolic engineering applications. *Appl Microbiol Biotechnol*, **83**, 199–208.

103. van den Brink, J., Canelas, A.B., van Gulik, W.M., Pronk, J.T., Heijnen, J.J., de Winde, J.H., and Daran-Lapujade, P. (2008) Dynamics of glycolytic regulation during adaptation of *Saccharomyces cerevisiae* to fermentative metabolism. *Appl Environ Microbiol*, **74**, 5710–5723.

104. Hadlich, F., Noack, S., and Wiechert, W. (2009) Translating biochemical network models between different kinetic formats. *Metab Eng*, **11**, 87–100.

105. Wahl, S.A., Noh, K., and Wiechert, W. (2008) 13C labeling experiments at metabolic nonstationary conditions: an exploratory study. *BMC Bioinformatics*, **9**, 152.

106. Glish, G.L. and Vachet, R.W. (2003) The basics of mass spectrometry in the twenty-first century. *Nat Rev Drug Discov*, **2**, 140–150.

107. Gross, J.H. (2011) In Gross, J.H. (ed.), *Mass Spectrometry*. Springer, Berlin and Heidelberg, pp. 651–684.

108. Halket, J.M., Waterman, D., Przyborowska, A.M., Patel, R.K.P., Fraser, P.D., and Bramley, P.M. (2005) Chemical derivatization and mass spectral libraries in metabolic profiling by GC/MS and LC/MS/MS. *J Exp Bot*, **56**, 219–243.

109. Bothwell, J.H. and Griffin, J.L. (2011) An introduction to biological nuclear magnetic resonance spectroscopy. *Biol Rev Camb Philos Soc*, **86**, 493–510.

110. McNair, H.M. and Miller, J.M. (2011) *Basic Gas Chromatography*. John Wiley & Sons, Hoboken, NJ.

111. Dauner, M. and Sauer, U. (2000) GC-MS analysis of amino acids rapidly provides rich information for isotopomer balancing. *Biotechnol Prog*, **16**, 642–649.

112. Fales, H.M. and Luukkainen, T. (1965) O-methyloximes as carbonyl derivatives in gas chromatography, mass spectrometry, and nuclear magnetic resonance. *Anal Chem*, **37**, 955–957.

113. Thenot, J.P. and Horning, E.C. (1972) Amino acid N-dimethylaminomethylene alkyl esters. New derivatives for GC and GC-MS studies. *Anal Lett*, **5**, 519–529.

114. Husek, P. (1991) Amino acid derivatization and analysis in five minutes. *FEBS Lett*, **280**, 354–356.

115. Hubschmann, H.-J. (2007) *Handbook of GC/MS: Fundamentals and Applications.* Wiley-VCH Verlag GmbH & Co. KGaA, Weinheim.

116. Jandera, P. and Henze, G. (2010) *Ullmann's Encyclopedia of Industrial Chemistry.* Wiley-VCH Verlag GmbH & Co. KGaA, Weinheim, pp. 85–138.

117. Erickson, B.E. (2000) HPLC and the ever popular LC/MS. *Anal Chem*, **72**, 711A–716A.

118. Cech, N.B. and Enke, C.G. (2001) Practical implications of some recent studies in electrospray ionization fundamentals. *Mass Spectrom Rev*, **20**, 362–387.

119. Matuszewski, B.K., Constanzer, M.L., and Chavez-Eng, C.M. (2003) Strategies for the assessment of matrix effect in quantitative bioanalytical methods based on HPLC-MS/MS. *Anal Chem*, **75**, 3019–3030.

120. Wittmann, C., Hans, M., and Heinzle, E. (2002) *In vivo* analysis of intracellular amino acid labelings by GC/MS. *Anal Biochem*, **307**, 379–382.

121. van Winden, W.A., van Dam, J.C., Ras, C., Kleijn, R.J., Vinke, J.L., van Gulik, W.M., and Heijnen, J.J. (2005) Metabolic-flux analysis of *Saccharomyces cerevisiae* CEN.PK113-7D based on mass isotopomer measurements of (13)C-labeled primary metabolites. *FEMS Yeast Res*, **5**, 559–568.

122. Noh, K., Gronke, K., Luo, B., Takors, R., Oldiges, M., and Wiechert, W. (2007) Metabolic flux analysis at ultra short time scale: isotopically non-stationary 13C labeling experiments. *J Biotechnol*, **129**, 249–267.

123. Toya, Y., Ishii, N., Hirasawa, T., Naba, M., Hirai, K., Sugawara, K., Igarashi, S., Shimizu, K., Tomita, M., and Soga, T. (2007) Direct measurement of isotopomer of intracellular metabolites using capillary electrophoresis time-of-flight mass spectrometry for efficient metabolic flux analysis. *J Chromatogr A*, **1159**, 134–141.

124. Antoniewicz, M.R., Kelleher, J.K., and Stephanopoulos, G. (2007) Accurate assessment of amino acid mass isotopomer distributions for metabolic flux analysis. *Anal Chem*, **79**, 7554–7559.

125. Kiefer, P., Nicolas, C., Letisse, F., and Portais, J.C. (2007) Determination of carbon labeling distribution of intracellular metabolites from single fragment ions by ion chromatography tandem mass spectrometry. *Anal Biochem*, **360**, 182–188.

126. Pingitore, F., Tang, Y., Kruppa, G.H., and Keasling, J.D. (2007) Analysis of amino acid isotopomers using FT-ICR MS. *Anal Chem*, **79**, 2483–2490.

127. Balci, M. (2005) *Basic ¹H- and ¹³C-NMR Spectroscopy.* Elsevier, Amsterdam.

128. London, R.E. (1988) 13C labeling in studies of metabolic regulation. *Prog NMR Spectr*, **20**, 337–383.

129. Szyperski, T. (1995) Biosynthetically directed fractional 13C-labeling of proteinogenic amino acids. An efficient analytical tool to investigate intermediary metabolism. *Eur J Biochem*, **232**, 433–448.

130. de Graaf, R.A., Rothman, D.L., and Behar, K.L. (2011) State of the art direct 13C and indirect 1H-[13C] NMR spectroscopy *in vivo.* A practical guide. *NMR Biomed,* **24**, 958–972.

131. Sonntag, K., Eggeling, L., De Graaf, A.A., and Sahm, H. (1993) Flux partitioning in the split pathway of lysine synthesis in *Corynebacterium glutamicum.* Quantification by 13C- and 1H-NMR spectroscopy. *Eur J Biochem,* **213**, 1325–1331.

132. Wendisch, V.F., de Graaf, A.A., and Sahm, H. (1997) Accurate determination of 13C enrichments in nonprotonated carbon atoms of isotopically enriched amino acids by 1H nuclear magnetic resonance. *Anal Biochem,* **245**, 196–202.

133. Ernst, R.R. and Anderson, W.A. (1966) Application of Fourier transform spectroscopy to magnetic resonance. *Rev Sci Instrum,* **37**, 93.

134. Eakin, R.T., Morgan, L.O., Gregg, C.T., and Matwiyoff, N.A. (1972) Carbon-13 nuclear magnetic resonance spectroscopy of living cells and their metabolism of a specifically labeled 13C substrate. *FEBS Lett,* **28**, 259–264.

135. Inbar, L., Kahana, Z.E., and Lapidot, A. (1985) Natural-abundance 13C nuclear magnetic resonance studies of regulation and overproduction of L-lysine by *Brevibacterium flavum. Eur J Biochem,* **149**, 601–607.

136. Walker, T.E., Han, C.H., Kollman, V.H., London, R.E., and Matwiyoff, N.A. (1982) 13C nuclear magnetic resonance studies of the biosynthesis by *Microbacterium ammoniaphilum* of L-glutamate selectively enriched with carbon-13. *J Biol Chem,* **257**, 1189–1195.

137. Wahl, A., El Massaoudi, M., Schipper, D., Wiechert, W., and Takors, R. (2004) Serial 13C-based flux analysis of an L-phenylalanine-producing *E. coli* strain using the sensor reactor. *Biotechnol Prog,* **20**, 706–714.

138. Krivdin, L.B. and Kalabin, G.A. (1989) Structural applications of one-bond carbon-carbon spin-spin coupling constants. *Prog NMR Spectr,* **21**, 293–448.

139. Emmerling, M., Dauner, M., Ponti, A., Fiaux, J., Hochuli, M., Szyperski, T., Wuthrich, K., Bailey, J.E., and Sauer, U. (2002) Metabolic flux responses to pyruvate kinase knockout in *Escherichia coli. J Bacteriol,* **184**, 152–164.

140. Dauner, M., Bailey, J.E., and Sauer, U. (2001) Metabolic flux analysis with a comprehensive isotopomer model in *Bacillus subtilis. Biotechnol Bioeng,* **76**, 144–156.

141. Dauner, M., Storni, T., and Sauer, U. (2001) *Bacillus subtilis* metabolism and energetics in carbon-limited and excess-carbon chemostat culture. *J Bacteriol,* **183**, 7308–7317.

142. Dauner, M., Sonderegger, M., Hochuli, M., Szyperski, T., Wuthrich, K., Hohmann, H.P., Sauer, U., and Bailey, J.E. (2002) Intracellular carbon fluxes in riboflavin-producing *Bacillus subtilis* during growth on two-carbon substrate mixtures. *Appl Environ Microbiol,* **68**, 1760–1771.

143. Massou, S., Nicolas, C., Letisse, F., and Portais, J.C. (2007) Application of 2D-TOCSY NMR to the measurement of specific ^{13}C-enrichments in complex mixtures of ^{13}C-labeled metabolites. *Metab Eng,* **9**, 252–257.

144. Christensen, B. and Nielsen, J. (2000) Metabolic network analysis of *Penicillium chrysogenum* using (13)C-labeled glucose. *Biotechnol Bioeng,* **68**, 652–659.

145. Calvin, M., Heidelberger, C., Reid, J., Tolbert, B., and Yankwich, P. (1949) *Isotopic Carbon.* John Wiley and Sons, New York.

146. Doudoroff, M. and Stanier, R.Y. (1959) Role of poly-beta-hydroxybutyric acid in the assimilation of organic carbon by bacteria. *Nature*, **183**, 1440–1442.

147. den Hollander, J.A., Ugurbil, K., Brown, T.R., Bednar, M., Redfield, C., and Shulman, R.G. (1986) Studies of anaerobic and aerobic glycolysis in *Saccharomyces cerevisiae. Biochemistry*, **25**, 203–211.

148. Beevers, H. and Gibbs, M. (1954) Participation of the oxidative pathway in yeast respiration. *Nature*, **173**, 640–641.

149. Yang, T.H., Heinzle, E., and Wittmann, C. (2005) Theoretical aspects of 13C metabolic flux analysis with sole quantification of carbon dioxide labeling. *Comput Biol Chem*, **29**, 121–133.

150. Yang, T.H., Wittmann, C., and Heinzle, E. (2006) Respirometric 13C flux analysis, Part I: design, construction and validation of a novel multiple reactor system using on-line membrane inlet mass spectrometry. *Metab Eng*, **8**, 417–431.

151. Hoon Yang, T., Wittmann, C., and Heinzle, E. (2006) Respirometric 13C flux analysis—Part II: *in vivo* flux estimation of lysine-producing *Corynebacterium glutamicum. Metab Eng*, **8**, 432–446.

152. Ishino, S., Kuga, T., Yamaguchi, K., Shirahata, K., and Araki, K. (1986) 13C NMR studies of histidine fermentation with a *Corynebacterium glutamicum* mutant. *Agric Biol Chem*, **50**, 307–310.

153. Walsh, K. and Koshland, D.E., Jr (1984) Determination of flux through the branch point of two metabolic cycles. The tricarboxylic acid cycle and the glyoxylate shunt. *J Biol Chem*, **259**, 9646–9654.

154. Rollin, C., Morgant, V., Guyonvarch, A., and Guerquin-Kern, J.L. (1995) 13C-NMR studies of *Corynebacterium melassecola* metabolic pathways. *Eur J Biochem*, **227**, 488–493.

155. Sauer, U. (2006) Metabolic networks in motion: 13C-based flux analysis. *Mol Syst Biol*, **2**, 62.

156. Sauer, U., Lasko, D.R., Fiaux, J., Hochuli, M., Glaser, R., Szyperski, T., Wuthrich, K., and Bailey, J.E. (1999) Metabolic flux ratio analysis of genetic and environmental modulations of *Escherichia coli* central carbon metabolism. *J Bacteriol*, **181**, 6679–6688.

157. Nanchen, A., Fuhrer, T., and Sauer, U. (2007) Determination of metabolic flux ratios from 13C-experiments and gas chromatography-mass spectrometry data: protocol and principles. *Methods Mol Biol*, **358**, 177–197.

158. Tannler, S., Decasper, S., and Sauer, U. (2008) Maintenance metabolism and carbon fluxes in Bacillus species. *Microb Cell Fact*, **7**, 19.

159. Nanchen, A., Schicker, A., Revelles, O., and Sauer, U. (2008) Cyclic AMP-dependent catabolite repression is the dominant control mechanism of metabolic fluxes under glucose limitation in *Escherichia coli. J Bacteriol*, **190**, 2323–2330.

160. Tannler, S., Fischer, E., Le Coq, D., Doan, T., Jamet, E., Sauer, U., and Aymerich, S. (2008) CcpN controls central carbon fluxes in *Bacillus subtilis. J Bacteriol*, **190**, 6178–6187.

161. Rantanen, A., Rousu, J., Jouhten, P., Zamboni, N., Maaheimo, H., and Ukkonen, E. (2008) An analytic and systematic framework for estimating metabolic flux ratios from 13C tracer experiments. *BMC Bioinformatics*, **9**, 266.

162. Zupke, C. and Stephanopoulos, G. (1994) Modeling of isotope distributions and intracellular fluxes in metabolic networks using atom mapping matrices. *Biotechnol Prog*, **10**, 489–498.

163. Marx, A., de Graaf, A.A., Wiechert, W., Eggeling, L., and Sahm, H. (1996) Determination of the fluxes in the central metabolism of *Corynebacterium glutamicum* by nuclear magnetic resonance spectroscopy combined with metabolite balancing. *Biotechnol Bioeng*, **49**, 111–129.

164. Christensen, B., Gombert, A.K., and Nielsen, J. (2002) Analysis of flux estimates based on (13)C-labelling experiments. *Eur J Biochem*, **269**, 2795–2800.

165. Dauner, M. (2000) Swiss Federal Institute of Technology, Zurich.

166. Schmidt, K., Carlsen, M., Nielsen, J., and Villadsen, J. (1997) Modeling isotopomer distributions in biochemical networks using isotopomer mapping matrices. *Biotechnol Bioeng*, **55**, 831–840.

167. Schmidt, K., Nielsen, J., and Villadsen, J. (1999) Quantitative analysis of metabolic fluxes in *Escherichia coli*, using two-dimensional NMR spectroscopy and complete isotopomer models. *J Biotechnol*, **71**, 175–189.

168. Wiechert, W., Mollney, M., Isermann, N., Wurzel, M., and de Graaf, A.A. (1999) Bidirectional reaction steps in metabolic networks: III. Explicit solution and analysis of isotopomer labeling systems. *Biotechnol Bioeng*, **66**, 69–85.

169. Wiechert, W. (2001) 13C metabolic flux analysis. *Metab Eng*, **3**, 195–206.

170. Antoniewicz, M.R., Kelleher, J.K., and Stephanopoulos, G. (2007) Elementary metabolite units (EMU): a novel framework for modeling isotopic distributions. *Metab Eng*, **9**, 68–86.

171. Iwatani, S., Yamada, Y., and Usuda, Y. (2008) Metabolic flux analysis in biotechnology processes. *Biotechnol Lett*, **30**, 791–799.

172. Young, J.D., Walther, J.L., Antoniewicz, M.R., Yoo, H., and Stephanopoulos, G. (2008) An elementary metabolite unit (EMU) based method of isotopically nonstationary flux analysis. *Biotechnol Bioeng*, **99**, 686–699.

173. David, H., Krogh, A.M., Roca, C., Akesson, M., and Nielsen, J. (2005) CreA influences the metabolic fluxes of *Aspergillus nidulans* during growth on glucose and xylose. *Microbiology*, **151**, 2209–2221.

174. Thykaer, J. and Nielsen, J. (2007) Evidence, through C13-labelling analysis, of phosphoketolase activity in fungi. *Process Biochem*, **42**, 1050–1055.

175. Wittmann, C. (2007) Fluxome analysis using GC-MS. *Microb Cell Fact*, **6**, 6.

176. Zamboni, N. and Sauer, U. (2009) Novel biological insights through metabolomics and (13)C-flux analysis. *Curr Opin Microbiol*, **12**, 553–558.

177. Kummel, A., Panke, S., and Heinemann, M. (2006) Putative regulatory sites unraveled by network-embedded thermodynamic analysis of metabolome data. *Mol Syst Biol*, **2**, 2006.0034.

178. Henry, C.S., Broadbelt, L.J., and Hatzimanikatis, V. (2007) Thermodynamics-based metabolic flux analysis. *Biophys J*, **92**, 1792–1805.

179. Hoppe, A., Hoffmann, S., and Holzhutter, H.G. (2007) Including metabolite concentrations into flux balance analysis: thermodynamic realizability as a constraint on flux distributions in metabolic networks. *BMC Syst Biol*, **1**, 23.

180. Beard, D.A. and Qian, H. (2007) Relationship between thermodynamic driving force and one-way fluxes in reversible processes. *PLoS ONE*, **2**, e144.

181. Wiechert, W. (2007) The thermodynamic meaning of metabolic exchange fluxes. *Biophys J*, **93**, 2255–2264.

182. Mavrovouniotis, M.L. (1990) Group contributions for estimating standard Gibbs energies of formation of biochemical compounds in aqueous solution. *Biotechnol Bioeng*, **36**, 1070–1082.

183. Mavrovouniotis, M.L. (1991) Estimation of standard Gibbs energy changes of biotransformations. *J Biol Chem*, **266**, 14440–14445.

184. Maskow, T. and von Stockar, U. (2005) How reliable are thermodynamic feasibility statements of biochemical pathways? *Biotechnol Bioeng*, **92**, 223–230.

185. Jankowski, M.D., Henry, C.S., Broadbelt, L.J., and Hatzimanikatis, V. (2008) Group contribution method for thermodynamic analysis of complex metabolic networks. *Biophys J*, **95**, 1487–1499.

186. Szyperski, T. (1998) ^{13}C-NMR, MS and metabolic flux balancing in biotechnology research. *Q Rev Biophys*, **31**, 41–106.

3

INTEGRATION OF "OMICS" DATA WITH GENOME-SCALE METABOLIC MODELS

Stephen Van Dien, Priti Pharkya, and Robin Osterhout

3.0 INTRODUCTION

Biological discovery and applied biotechnology are constantly challenged by the complexity of living cells. The existence of unknown factors and poorly understood processes often leads to unexpected results, which translates to inefficiencies in both discovery research and product development. With the advent of high-throughput technologies, such as automated DNA sequencing, genome-wide expression analysis, proteomics, and high-throughput screening, the number of unknown cellular components is being reduced. The identification of these factors is, however, just the first step. In order for these data to provide tangible benefits in the scientific, medical, and industrial communities, technology must be established to interpret this information in the context of the entire biological system (1). Only then can we begin to utilize such knowledge to predict biological functions and responses, and thus ultimately alleviate the challenges that the complex system presents.

As the common denominator of all cellular functions, metabolism offers the best place from which to base the development of *in silico* models used to evaluate high-throughput "omics" data. Through the utilization of a vast repertoire of enzymatic reactions and transport processes, unicellular and multicellular organisms can process and convert thousands of organic compounds into the various biomolecules necessary to support their existence. In

Engineering Complex Phenotypes in Industrial Strains, First Edition. Edited by Ranjan Patnaik.
© 2013 John Wiley & Sons, Inc. Published 2013 by John Wiley & Sons, Inc.

switchboard-like fashion an organism directs the distribution and processing of metabolites throughout its extensive map of pathways. The combination of metabolic models with an experimental research platform can have a powerful impact on the biotechnology industry as a whole. Various technological advances at the DNA sequence, transcript, and protein levels have accelerated our ability to characterize and quantify the components of a biocatalyst's metabolic machinery. In addition, the continuous development of recombinant DNA techniques has made it increasingly possible to rationally manipulate the genetic content of virtually any candidate production organism. Metabolic models provide a means to capture this data in an organized manner, and translate it into phenotypic behavior through simulations. In addition, availability of new data will drive iterative model development. These *in silico* models therefore serve as the most concise representation of the biology and metabolism of a microorganism. As such they can become the focal point for the integrative analysis of vast amounts of experimental data and a central resource to design experiments and drive research programs (1,2).

3.1   GENOME-SCALE METABOLIC NETWORKS

Genome-scale models provide a framework to organize genome sequence information and interpret data in the context of cell metabolism. In order to analyze, interpret, and predict cellular behavior using metabolic simulations, each individual link in a biochemical network must be described, normally with a rate equation that requires a number of kinetic parameters. Unfortunately, it is currently not possible to formulate this level of description for cellular processes on a genome scale. In the absence of kinetic information, it is still possible to assess the theoretical capabilities of integrated cellular processes by using a data-driven, constraint-based approach (3). Rather than attempting to calculate and predict exactly what a metabolic network does, we are able to narrow the range of possible phenotypes that a metabolic system can display based on the successive imposition of governing physico-chemical constraints. Thus, instead of calculating an exact phenotypic "solution," we can determine the feasible set of phenotypic solutions in which the cell can operate (illustrated in Figure 3.2). Optimization procedures are then applied to calculate the "best" solution within the allowable range based on a particular objective function, such as maximizing growth yield (4–6). If the network has evolved to produce the "best" or optimal function, then agreement is reached between experimentally determined behavior and the *in silico* computations.

Construction of genome-scale metabolic models begins with the metabolic reconstruction, in which genes are linked to proteins and ultimately to the reactions they catalyze (7). Homology searches are used to assign putative function to each gene. Proteins are then connected to reactions using "AND/ OR" logic; for example, several proteins may need to form a complex in order to catalyze a single reaction, one enzyme may perform multiple reactions, or several isozymes may catalyze the same reaction. Maintaining the correct model structure in this dimension is crucial for the interpretation of omics data. These steps can often be automated using computational tools, such as the SimPheny platform (Genomatica, Inc., San Diego, CA, http://www.genomatica.com). Due to incomplete genome sequences, genes of unknown function, and protein sequences diverging to the extent that they no longer can be found by homology searching, it is almost certain that the initial metabolic reconstruction will be incomplete. Visualization of the pathways on maps can highlight where information is complete and where there are knowledge gaps (8). Such gaps result in dead-end metabolites, orphan genes, or even missing pathways. Gaps in the network can be filled by physiological knowledge. For example, if only one or two steps are missing in an essential biosynthetic pathway, it is likely that the pathway is present. Research can then be focused on finding the genes encoding the missing steps, if desired (9). Use of the models in conjunction with experimental data can also find previously unknown or poorly characterized pathways, even when they are not essential for growth on standard laboratory media. Intracellular metabolite measurements ("metabolomics" data) can find metabolites that are not present in the model (10). These metabolites can be linked to the rest of the metabolic network using gap-filling algorithms (11), or often just by visual inspection. In general, model development is an iterative process (Figure 3.1). Models are

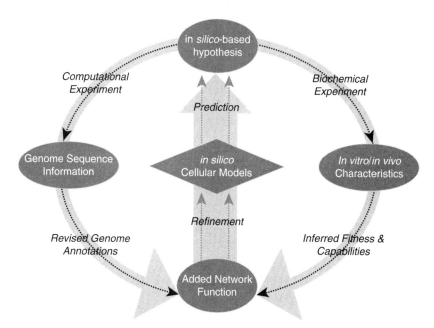

FIGURE 3.1. The *in silico* model is used to generate hypotheses testable through new experimental design and through further analysis of existing experimental/bioinformatics data. The model leads to the design of experiments to enhance our understanding of the organism, leading to refinements in the model and the notion of iterative model development to accelerate discovery.

constructed based on the genome sequence and limited experimental data. The model can then identify poorly characterized regions of metabolism, which serve as the focus for further research. This work may then identify inconsistencies, which are reconciled to improve the model. Although this will improve model predictions in certain cases, it may also create new questions, and thus the cycle starts over. In summary, the model drives research effort, and the research drives model improvement.

Use of growth-phenotyping data, such as that generated by Biolog (Hayward, CA, http://www.biolog.com), is a relatively inexpensive way to leverage models for a systems approach to metabolic network discovery and improving gene annotations. In a recent study with *Bacillus subtilis*, phenotype microarray (PM) data were used to add 75 reactions to the genome-scale model, which were essential for growth on certain substrates that tested positive in the array (12). Forty-nine of these reactions are for metabolite transport, adding to the large number (191) of transporters already in the model based on genetic and biochemical evidence. For some of the substrates, addition of the appropriate transporter was not sufficient to allow growth. Catabolic reactions also had to be added to link certain carbon sources, such as D-malate, L-arabitol, and dulcitol, to pathways that are already present in the model. The authors then used bioinformatic approaches to find candidate genes that may encode these reactions. Although experimental validation is required, such a model-driven approach provides targets and can save time by focusing future research in directions that are most likely to succeed. Reed and coworkers have demonstrated this methodology for improved annotation of *Escherichia coli* gene functions (9). As with the *Bacillus* study, PMs from Biolog were used to find discrepancies between observed phenotype and model predictions, and putative reactions were added to the model. Genes encoding these reactions were identified by a combination of homology searches, gene expression by microarray analysis and reverse transcriptase-polymerase chain reaction (RT-PCR), enzyme activity assays, and finally confirmed by growth phenotype of deletion mutants on the substrate of interest. A clear example is the utilization of D-malate, for which the transporter (*dctA*), dehydrogenase (*yeaU*), and a regulator (*yeaT*) were identified.

3.2 CONSTRAINT-BASED MODELING THEORY

The core principle underlying the constraint-based approach lies in the balance equations imposed by the stoichiometry of the reactions. Basically, this represents mass, energy, and redox balance constraints. The mathematical formalism of stoichiometric modeling is well developed (13) and has been used extensively in the field of metabolic engineering under the terms flux balance analysis (FBA) and metabolic pathway analysis (MPA) (5,6,14,15,15,16) (see Chapter 2). It is based on the application of a pseudo-steady state hypothesis (16) to the mass balance of metabolites, yielding the equation

$$\mathbf{S} \cdot \mathbf{v} = 0.$$

where $\mathbf{S}$ is the stoichiometric matrix and $\mathbf{v}$ the vector of reaction rates. The number of reactions is almost always greater than the number of metabolites, resulting in an underdetermined system of equations. Thus, instead of a unique solution, we end up with a feasible solution space as shown in Figure 3.2. It is

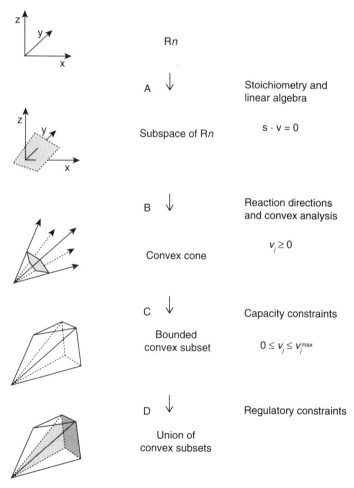

FIGURE 3.2. Schematic view of constraint-based modeling. The successive imposition of governing constraints, including the (A) stoichiometric, (B) thermodynamic, and (C) enzyme capacity constraints, reduces the size of the feasible set (shown as bounded space within dotted lines and shaded planes) and is represented through the incorporation of a set of mathematical statements. In the limiting case where all constraints on the metabolic network are known (D), for example, enzyme kinetics and gene regulation, the feasible set may be reduced to a single point.

this space to which we apply our successive constraints and optimization procedures.

Such a constraint-based approach provides a basis for understanding the structure and function of biochemical networks (5,6,14,15,16). Beginning with the solution to the above equation, we then add capacity limitations to account for the maximum flux through enzymatic reactions. We can further constrain the solution space by adding experimental data: extracellular fluxes obtained from fermentation process data; qualitative constraints based on known metabolite-mediated regulatory interactions; qualitative constraints from gene expression data; and the assignment of flux directions based on thermodynamic constraints imposed by metabolite concentrations. Each additional data input provides new information that can be used to reduce the range of feasible flux distributions and phenotypes that a metabolic network can display. The latter two types of constraints indicated above, derived from gene expression and metabolomics data, is the subject of this review.

3.3   CURRENT ANALYSIS OF OMICS DATA

Genomics data form the basis of model development, as already discussed. In the post-genomic era, a variety of omics technologies have been developed to collect high-throughput data characterizing levels of the central dogma leading from genotype to phenotype. This includes transcriptomics (microarrays), proteomics, metabolomics, fluxomics, and phenomics (such as the Biolog PMs). Systems biology seeks to utilize this information to map the genotype–phenotype relationship, and ultimately understand the complex behavior of living cells (17). However, most methods commonly used for the analysis of such data fall short of this objective, limiting the scope of the analysis to only one aspect of cell physiology such as gene regulation circuits. On the other hand, *in silico* models allow for the holistic investigation of biological systems, capturing the activity of multiple gene products working together in a globally orchestrated fashion. Therefore, these models provide an ideal platform for the analysis and interpretation of omics data to drive both scientific discovery and biotechnology applications (18).

Since the pioneering work in the mid-1990s, gene expression microarrays have become ubiquitous. Virtually every major commercial research effort focused on understanding cellular responses in medical and industrial biotechnology is implementing whole-genome expression profiling along with a suite of other experimental technologies. Concomitant with the improvement in the technology for creating gene chips and performing microarray experiments, a variety of statistical tools have been developed to analyze and interpret the large quantity of data generated from such experiments. A major challenge recognized early on was how to determine if changes in expression between two conditions were significant. The simplest method is to apply a heuristic such as a twofold expression change compared with baseline (19). Higher confidence results were obtained using *t*-tests assuming a Gaussian distribution (20), or nonparametric methods that do not assume a normal distribution (21). The problem becomes more complex when comparing multiple samples. Methods such as singular value decomposition (22,23), self-organizing maps (24), or hierarchical clustering (25) are used to group genes based on expression patterns over time or across different conditions. However, the utility of these approaches hinges on the assumption that genes with similar expression behavior are likely to be related functionally, without consideration for the biological context of the genes (26). Integration of these data with other omics data sets and a biological model, particularly a constraints-based model, can provide further insight on a systems level (27). This is not a substitute for the above methods, but rather a complement.

Identification and measurement of intracellular metabolites gets us a step closer to phenotypic characterization, but still can be linked to genotype through the known metabolic reactions connecting these metabolites. Thus, in one sense, such metabolomics data are ideal for investigating the genotype–phenotype relationship. Metabolomics research has lagged behind that of transcriptomics and proteomics, primarily due to the wide variety of chemistries involved and thus the need for multiple measurement techniques. Nonetheless, metabolomics is gaining traction in both industrial and academic research groups, and significant improvements have been made in both the number of metabolites that can be identified and the accuracy with which intracellular concentrations can be measured (10,28–30). Many of the same statistical techniques applied to microarrays were used with metabolomics data to look at trending in profiles among different data sets (31–35). Such work has proven useful in discovering biomarkers and identifying strains (36), but provides no biological insight. On the other hand, analysis of metabolomics data in conjunction with metabolic models will lead to a better understanding of the metabolic processes that drive cell function, and will utilize these data sets to their full potential (37–39).

3.4 NEW APPROACHES TO DEVELOPING MODEL CONSTRAINTS

To improve the predictive capability of metabolic models without the use of experimental data, new rules must be implemented to restrict the feasible flux space. One such set of rules is thermodynamic constraints. To a first approximation, thermodynamics have been traditionally incorporated into models by specifying each reaction as either reversible or irreversible. Often such restrictions are chosen somewhat arbitrarily, based on the way they are written in biochemistry texts or internet reaction databases, without much true regard for the ΔG values. The latest genome-scale *E. coli* model is more rigorous in its reaction direction assignments (40). Gibbs energies of formation are calculated for each metabolite using the group contribution method (41,42), and used to calculate the possible ΔG range for each reaction given a typical physiological range for metabolite concentrations. A related field, energy balance analysis, was developed to eliminate energy-generating cycles using loop laws akin to electric circuits (43). Thus, even in a set of cyclic equations where each reaction individually could proceed in the appropriate direction, the second law of thermodynamics prohibits operation of a cycle unless one of the reactions has external energy input such as adenosine triphosphate (ATP). Although such infeasible cycles could often be detected by inspection and corrected by making one reaction irreversible, energy balance analysis formalized the process.

More recently, two new approaches have been developed to incorporate thermodynamic feasibility constraints in metabolic flux analysis: network-embedded thermodynamic analysis (NET analysis) (44) and thermodynamics-based metabolic flux analysis (TMFA) (45). In addition, both of these methods can incorporate metabolomics data (intracellular metabolite concentration profiles) to further constrain the simulations. NET analysis employs a nonlinear optimization algorithm to calculate feasible ranges of metabolite concentrations and Gibbs energies of reaction based on the topology of a metabolic network and observed metabolite profiles (44). NET analysis can be applied to evaluate the quality of metabolomics data, identify putative regulatory sites, and predict feasible concentration ranges of unmeasured metabolites. Inputs to the optimization algorithm include a stoichiometric matrix of the network, a direction of flux for each reaction, the Gibbs energy of formation $\Delta_f G$ of metabolites, and concentration constraints. Flux directions can be derived from experimental measurements (c.f. [13]C flux measurements, see Chapter 2), from metabolic flux analysis predictions, or from preexisting knowledge of the network. Concentration limits can vary over a wide range (0.001–10 mM) or can be narrowly constrained by metabolomics data. Error associated with thermodynamic parameters and concentration constraints is also incorporated into the optimization framework. TMFA is a linear optimization method recently developed for incorporating thermodynamic data directly into the metabolic flux analysis framework (45). Unlike NET analysis, TMFA does not require preexisting knowledge of flux directions. Rather, the method integrates thermodynamic and mass-balance constraints to ensure that predicted flux distributions are thermodynamically feasible (Figure 3.3). TMFA can be applied to predict flux and Gibbs energy ranges of each reaction, and also a feasible concentration range of each metabolite. While NET analysis is formulated as a nonlinear optimization, TMFA as described by Henry et al. (45) is a linear problem. As such, TMFA is guaranteed to converge on a global optimum solution. A minor limitation is that a handful of the NET analysis constraints cannot be implemented within the linear TMFA framework (e.g., the ratios of the NAD or NADP cofactors and the adenylate energy charge). Both techniques have been developed and applied to small metabolomics data

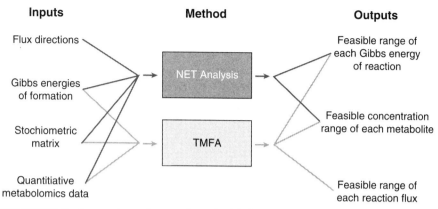

FIGURE 3.3. Comparison of network-embedded thermodynamic analysis (NET) and thermodynamics-based metabolic flux analysis (TMFA). Inputs and outputs to NET analysis and TMFA are color coded in red and blue, respectively. *(See insert for color representation of the figure.)*

sets. The true potential of metabolomics data for understanding metabolic behavior awaits to be seen, when these methods are applied to measurements of a hundred or more metabolites.

Another new development is the incorporation of regulatory constraints into the flux balance analysis (FBA) framework (rFBA approach). Many regulatory rules have been elucidated for common model organisms like *E. coli* and *Saccharomyces cerevisiae*, and such regulatory systems will prevent certain flux distributions from occurring under a given environment. For example, pyruvate formate lyase is not active in the presence of oxygen (46), and acetate is not taken up when glucose is present (47). Therefore, regulatory rules can constrain the metabolic network and sharpen model predictions. The most straightforward way to implement regulatory constraints is using a Boolean representation (48). For example, gene repression can be simulated by the following logic: if metabolite A exists in the cell (either due to a transporter or reaction producing it having positive flux), then reaction X is constrained to zero. A transient process, such as a batch culture, is simulated by running the metabolic/regulatory model successively over small time intervals. The output of one time step, including regulatory signals, provides the input to the next step. Using such a procedure, batch growth and metabolite secretion by *E. coli* were accurately predicted in a variety of conditions, including diauxic growth (48). A total of 104 regulatory genes were then incorporated into the genome-scale *E. coli* model, controlling expression of 479 out of 906 of the metabolic genes. Simulations were performed and compared with the measured growth phenotypes in over 13,000 combinations of single gene deletions and environmental conditions. The metabolism-only model correctly predicted 65% of the phenotypes, while the combined metabolism/regulatory model was accurate 79% of the time (49). The reason for this difference is that the regulatory model is more constrained and that some flux distributions feasible based on the stoichiometry alone become infeasible once the regulatory rules eliminate possible solutions.

Models can also be used in conjunction with data to elucidate unknown regulatory rules. Data from genome-wide chromatin immunoprecipitation experiments were used to find regulatory behavior that was not predicted by a metabolic/regulatory model of *S. cerevisiae* (50). Putative regulatory interactions that would match the experimental data if applied were then identified. A new approach termed EGRIN (Environmental Gene/Regulatory Network) can reconstruct an entire regulatory network based on data contained in hundreds of carefully designed microarray experiments conducted under a wide variety of environmental conditions. This technique makes use of two computational tools: cMONKEY (51) and the Inferelator (52). cMonkey uses a biclustering methodology (clustering both by genes and conditions) to identify putative regulons. As an input it takes gene expression data in conjunction with upstream regulatory sequences and relationship information such as metabolic pathway (functional) associations or protein interaction information. The Inferelator then predicts a set of regulatory interactions for each bicluster. It accomplishes this by identifying relationships between factors, including both external environmental signals and mRNA expression levels of predicted transcription factors, and the expression levels of clusters of genes. The use of time course data can help elucidate directionality of the interactions. When applied to all the biclusters identified by cMonkey, the result is a predicted regulatory network, or EGRIN.

Finally, FBA has been combined with regulatory Boolean logic and ordinary differential equations to model the dynamic behavior of metabolic, regulatory, and signaling networks (53,54). This integrated FBA approach, iFBA, was used to create a model of *E. coli* that combined the central metabolic model incorporating the transcriptional regulation and the ordinary-differential equation (ODE)-based model of carbohydrate uptake (53). The advantages of this model are that it contains a much greater level of detail for regulatory activities and events than the rFBA approach. It can also account for enzymes such as adenylate cyclase which do not participate directly in metabolism but are critical because of their role in other activities such as signal transduction. The merits of the iFBA approach over the ODE approach are that it helps to understand the global effects of a dynamic change because of its ability to calculate a flux distribution for an entire network with only a few additional parameters. The predictions of an integrated model were compared with those of the individual models (rFBA based and ODE based) to predict the phenotypes of single gene perturbations for diauxic growth on glucose/lactose and glucose/glucose-6-phosphate and were shown to be more accurate than either approach in several cases. For example, iFBA was able to predict the dynamic behavior of three metabolites and three transporters inadequately predicted by rFBA. It was also able to predict more accurate phenotypes than the ODE model for 85 out of 334 single gene perturbation simulations (53).

3.5 USE OF GENE EXPRESSION DATA IN METABOLIC MODELS

Integration of transcriptomics data into the constraints-based modeling framework has presented challenges due to the general lack of correlation between gene expression and metabolic phenotype (55). For example, cases have been found where fluxes increase more than 10-fold without corresponding expression changes in the genes encoding the reactions (56). For reversible reactions, flux direction can reverse rather rapidly upon condition changes even without any change in the gene expression level. The problem is further complicated by situations where multiple genes are needed to encode a multi-subunit enzyme, or different enzymes can catalyze the same reaction. It is not uncommon to find situations where one isoenzyme is induced while the other is repressed; for example, different fumarase genes are used under aerobic and anaerobic conditions (57). Taken together, these issues indicate that one cannot draw general conclusions about fluxes through metabolic reactions based on the expression of corresponding genes.

Typically, models have therefore been used as scaffolds to visualize the data and interpret it in the context of the metabolism of the whole system rather than using rigorous mathematical approaches to constrain the model using expression data. For example, microarray data was also used to investigate the possible causes of tolerance to furfural (58), a toxic compound in biomass hydrolysates. The expression data were visualized and analyzed with SimPheny (Genomatica, San Diego, CA), ArrayStar (Arraystar, Rockville, MD), and Network Component Analysis (NCA) (59). By overlaying the expression data of the furfural tolerant strain and the control strain on the metabolic pathways, it was clear that furfural increased the expression of several genes associated with the assimilation of sulfur into amino acids, primarily cysteine and methionine (Figure 3.4). Sulfur is supplied as sulfate in AM1 medium, the growth medium used in the study, and must be reduced to the level of hydrogen sulfide for incorporation, an energy-intensive reaction requiring four molecules of NADPH. The furfural-induced increase in the expression of these genes was in sharp contrast to the decreased expression of several other biosynthetic genes. An increase in tolerance was indeed observed when the medium was supplemented with cysteine, methionine, and other reduced sulfur sources, such as thiosulfate. No response was observed upon supplementation with taurine, a sulfur source that requires three molecules of NADPH for assimilation (Figure 3.4). Overexpression of the membrane-bound transhydrogenase, encoded by *pntAB*, also increased tolerance. All of these results suggested a mechanism for growth inhibition by furfural. When furfural is present in the culture it can be metabolized by *yqhD*, the high-affinity alcohol dehydrogenase with a very low binding constant K_m for

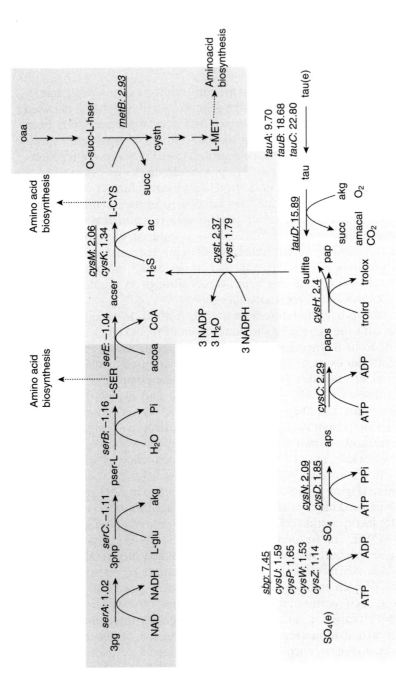

FIGURE 3.4. Pathways for sulfate assimilation, taurine metabolism, serine and cysteine biosynthesis and their connection with methionine biosynthesis. Fold-change gene expression levels are shown at each metabolic step, obtained from microarray data collected as described in ref. (54). Positive numbers show an upregulation of the gene when exposed to furfural, whereas negative numbers show a downregulation, and those that were upregulated more than twofold have been underlined.

NADPH (60), thereby depleting the cell of NADPH needed for biosynthesis. In this study, the effect of a toxic compound on the metabolism of *E. coli* was clearly elucidated by accounting for the overall metabolism of the organism.

In spite of the limitations of using the gene expression data for predicting metabolic phenotypes, there are clear examples where transcriptome data have provided insight into metabolic fluxes. One such study (55) examined expression data from batch and chemostat cultivations of *S. cerevisiae* on glucose (61), in conjunction with flux balance predictions from a genome-scale model of this organism (62). Using growth as the objective function, the model accurately predicted chemostat behavior but did not predict the reduced biomass yield and increased byproduct formation seen in the batch culture relative to the chemostat. The authors of this study then determined which genes had no evidence of expression (in biological triplicates as well as in replicate probes on the array), and constrained reactions associated with those genes to have zero flux (55). This resulted in the removal of 6 reactions for the chemostat case and 97 for the aerobic batch fermentation. Adding these constraints had little effect on the chemostat predictions, but clearly improved model performance for the batch culture. In contrast to the unconstrained model, the model constrained with gene expression data showed quantitative agreement for biomass, ethanol, and glycerol yield (Figure 3.5). Furthermore, the constrained model improved predictions of key branch point fluxes, known from ^{13}C-labeling experiments (see Chapter 2) (63). Expression data may also

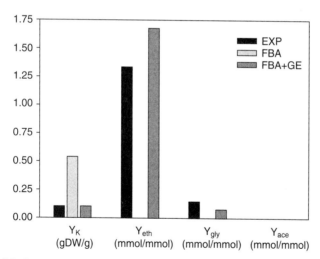

FIGURE 3.5. Improved prediction of biomass and product yields by using gene expression data in constraint-based modeling. Yields of biomass (Y_x), ethanol (Y_{eth}), glycerol (Y_{gly}), and acetate (Y_{ace}) in an aerobic batch cultivation of *S. cerevisiae* determined by experiment (EXP, black bars), standard flux balance analysis (FBA, light gray bars), and flux balance analysis combined with additional constraints from gene expression data (FBA+GE, medium gray bars). Reprinted from *Metabolic Engineering* (55), with permission from Elsevier.

be used to discriminate between alternate optimal solutions predicted by constraint-based modeling, particularly when these alternate solutions are distinguished primarily by isoenzymes and parallel pathways. Using the same expression data set (61), the authors showed how two alternate model solutions for aerobic chemostat growth differ drastically in their level of agreement with transcriptional upregulation and downregulation compared with an anaerobic case (55).

Patil and Nielsen integrated microarray data with network topology information as determined by a genome-scale metabolic model to predict the cellular response to perturbations (64). The premise behind the study was that the changes in individual gene expression levels in response to a perturbation are small and are not identified using standard statistical methods or clustering algorithms. Using genome-scale models, however, it is possible to identify patterns in the network that show a common transcriptional response. The authors (64) developed an algorithm that identifies a set of reporter metabolites (metabolites around which the most significant transcriptional changes occur) and a set of genes with significant and coordinated response to perturbations (Figure 3.6).

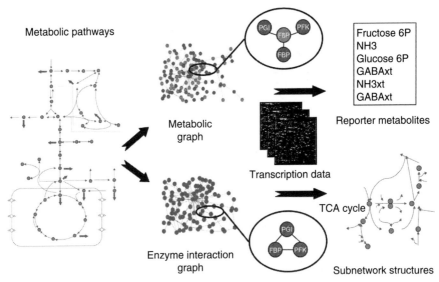

FIGURE 3.6. Illustration of the proposed algorithm for identifying reporter metabolites and subnetwork structures signifying transcriptionally regulated modules. A metabolic network is converted to metabolic and enzyme-interaction graph representations. Gene expression data from a particular experiment then are used to identify highly regulated metabolites (reporter metabolites) and significantly correlated subnetworks in the enzyme-interaction graph. TCA, tricarboxylic acid cycle; PGI, phosphoglucose isomerase; PFK, phosphofructokinase; FBP, fructose bisphosphatase. Reproduced from Reference (64), Copyright 2005 National Academy of Sciences, USA. *(See insert for color representation of the figure.)*

The same algorithm was used to analyze data from *S. cerevisiae* to identify reporter metabolites and the corresponding subnetworks in three cases: (i) deletion of a gene, (ii) deletion of a regulatory protein, and (iii) change in the environment of a cell. For example, the transcription data from a wild-type strain of *S. cerevisiae* was compared with a deletion mutant of the gene GDH1, encoding for NADPH-dependent glutamate dehydrogenase and involved in ammonia assimilation. The genome-scale metabolic model of *S. cerevisiae* was used to generate the metabolic and the reaction–interaction graphs. Using the algorithm, several key metabolites were identified including the three sugar phosphates: glucose-6-phosphate, sedoheptulose-7-phosphate, and fructose-6-phosphate. These three metabolites represent branch points between the Embden–Meyerhof–Parnas pathway and the pentose phosphate pathway. The deletion of GDH1 corresponds with a reduction of the growth-related requirement of NADPH of the cell by about 40%, and therefore less flux needs to be routed via the pentose phosphate pathway, the primary source of NADPH in *S. cerevisiae*. A high-scoring subnetwork of 34 genes was found, 10 of which involved NADH/NADPH, demonstrating the effects of GDH1 deletion on the redox metabolism. Two key nodes of metabolism were represented in this network: (i) the glycolysis-pentose phosphate node that is controlled by the requirement for NADPH and (ii) the alpha-ketoglutarate node. It has been shown that the level of alpha-ketoglutarate is increased in a ΔGDH1 mutant.

3.6   USE OF METABOLOMICS DATA IN METABOLIC MODELS: TMFA EXAMPLE

Advances in the field of metabolomics are enabling high-throughput and high-precision detection and quantification of metabolite concentrations. The resulting large-scale quantitative data sets can be evaluated in the context of *in silico* models to generate new insights into metabolism.

Several recently developed computational methods, including NET analysis and TMFA, directly incorporate metabolite concentration data into the constraint-based modeling framework (44,45,65). These algorithms integrate mass-balance constraints with thermodynamic principles by coupling flux directionality to the second law of thermodynamics, wherein reactions with a positive flux must have a negative Gibbs energy of reaction ($\Delta_{rxn}G < 0$) and reactions with a negative flux must have a positive Gibbs energy of reaction ($\Delta_{rxn}G > 0$). A schematic of TMFA analysis is shown in Figure 3.7. Inputs to the optimization include the genome-scale model (represented by the stoichiometric matrix of the network), the Gibbs energy of formation of metabolites ($\Delta_f G$), and concentration constraints. Gibbs energies of formation can be measured experimentally or calculated using group contribution methods (42,66). Metabolite concentrations can be allowed to vary over a physiologically relevant range, or can be narrowly constrained by experimental data. By

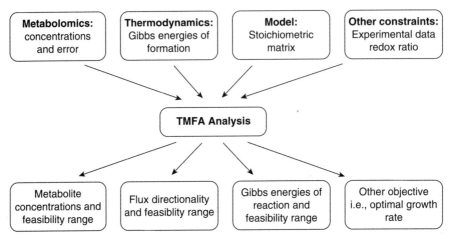

FIGURE 3.7. Schematic of thermodynamics-based metabolic flux analysis (TMFA). Inputs include available experimental and thermodynamic data and the stoichiometric matrix. Outputs include numerous objectives including thermodynamically feasible metabolite concentration, reaction flux, and Gibbs energy of reaction ranges.

integrating mass-balance and thermodynamic constraints with metabolomics data, thermodynamics-based flux analysis methods enable calculation of thermodynamically feasible solutions for a range of properties including flux, Gibbs energy of reaction, cofactor ratios, and concentration ranges of unmeasured metabolites.

Constraint-based methods that utilize thermodynamics and metabolomics data have numerous applications. The calculated *in vivo* Gibbs energies of reaction ($\Delta_{rxn}G$) are useful for identifying thermodynamic bottlenecks and putative regulatory sites. A reaction may function as a thermodynamic bottleneck, meaning it has low driving force, if the $\Delta_{rxn}G$ is constrained to operate very close to zero. In this case, small fluctuations in concentration can drive the flux through that reaction to zero or change the direction of flux (67,68). Knowledge of such bottlenecks can then trigger the search for bypass pathways, including novel pathways or existing pathways whose physiological significance is not yet appreciated. Alternatively, reactions that operate far from equilibrium (i.e., the $\Delta_{rxn}G$ range does not span 0) are more likely to serve as a regulatory control points for the pathways in which they participate (69,70). In addition to identifying potential thermodynamic bottlenecks and regulatory control points, thermodynamics-based flux analysis methods can be applied to evaluate the quality of metabolomics data and predict feasible concentration ranges of unmeasured metabolites (44). Other applications include improving the accuracy of new metabolic reconstructions (40), assessing reaction reversibility (71), and evaluating the feasibility of biodegradation reactions (66).

A recent study by Bennett and coworkers (72) successfully demonstrated how quantitative metabolomics data, in conjunction with *in silico* modeling methods, can be used to generate systems-level insights into metabolism. In this work, the authors introduced a novel ^{13}C isotope ratio-based method for precisely determining absolute metabolite concentrations of a large number of samples in a single experiment. The method was applied to generate high-quality quantitative measurements of >100 *E. coli* intracellular metabolite concentrations during aerobic growth on different carbon sources (glucose, acetate, and glycerol). The metabolome during growth on all carbon substrates was dominated by a handful of abundant compound classes: amino acids (49%), nucleotides (15%), central carbon intermediates (15%), and redox cofactors (9%). The detected concentrations ranged from 0.1 µM to 100 mM. Glutamate was the most abundant with an intracellular concentration of 100 mM, comprising 40% of the total molar concentration. A majority of metabolites were present at significantly different levels during growth on different substrates, and the applied methods enabled detection of small changes in concentration.

The Bennett et al. data set was evaluated in the context of the genome-scale *E. coli* metabolic model iJR904 (11) using TMFA. Metabolite concentrations, associated measurement errors, and thermodynamic properties were incorporated into the systems-level analysis. The optimal or near optimal growth solutions predicted by the iJR904 model satisfied the thermodynamically feasible constraints set by the data, validating the quality of both the data set and the metabolic model. Additionally, the *in vivo* free energy ranges of 25% of the known reactions in *E. coli* metabolism were calculated during growth on each carbon source. The authors found that over two-thirds of reactions are strongly forward-driven, with a ΔG less than -10 kJ/mol. Many of the reactions with free energies near equilibrium were observed in lower glycolysis, rendering these reactions more sensitive to fluctuations in metabolite concentrations and thus allowing these reactions to switch flux directions in response to different growth conditions.

The metabolomics data set was also evaluated in the context of enzyme kinetics by comparing the observed concentration of each metabolite to the K_M value of the enzymes that utilize the compound as a substrate. K_M values were extracted from the BRENDA database (http://www.brenda-enzymes.info). In a majority of cases (83%), the metabolite concentration exceeded K_M values. The measured concentrations of 59% of the metabolites were at least 10-fold higher than their K_M values, indicating saturation at their corresponding enzyme active sites. The substrates ATP and NAD^+ were nearly always saturating, whereas NADPH was not. Most glycolytic intermediates were present at saturating concentrations, indicating that other control mechanisms such as enzyme inhibition, activation, and availability regulate metabolic flux through glycolysis. Metabolites involved in degradation reactions, on the other hand, were typically nonsaturating, indicating that substrate availability plays a key role in regulating flux through these pathways. Substrates of central metabolic, tricarboxylic acid (TCA) cycle, and pentose phosphate pathway enzymes were generally present at levels close to their K_M values. This observation, along with the near-equilibrium *in vivo* free energies of many of the enzymes, is consistent with the bidirectional nature of the central carbon metabolic pathways. These findings also support the role of substrate availability in regulating the flux of bidirectional, but not unidirectional, reactions.

The Bennett et al. study nicely demonstrates the assimilation of high-quality quantitative metabolite concentration measurements with experimental data from the literature and public databases, and analysis of this data in the context of a metabolic model, which itself is curated and validated using experimental data. As metabolomics technologies become increasingly high-throughput, precise, and reproducible, quantitative metabolite profiling is certain to provide fundamental insights into cell physiology and the roles of thermodynamics and enzyme kinetics in regulating metabolic flux. Knowledge of how fluxes are regulated will be critical for complex metabolic engineering applications.

3.7 INTEGRATION OF MULTIPLE OMICS DATA SETS

Complex biological phenotypes can arise from the interplay of regulation at various levels of the central dogma. This is often reflected as contradictions between different types of omics data sets. It is important to embrace these differences to better understand how these different modes of control are integrated to give the observed phenotype. As the collection of omics data sets becomes more routine and the data more reliable, several researchers have recently taken the step to assimilate quantitative data sets from different omics levels, with the ultimate goal of understanding the complex relationship between genotype and phenotype. Ultimately, these relationships must be understood in order to apply rational engineering manipulations to manipulate metabolism as a whole.

Building on their work to identify reporter metabolites using transcriptome data (64), Cakir and coworkers (73) developed a hypothesis-driven algorithm to integrate metabolome data with metabolic models to detect reporter reactions. These are reactions that have significant changes in the levels of metabolites surrounding them following a genetic or environmental perturbation. The results of the metabolome study were then combined with transcriptome data to understand the mode of regulation. A graph theoretical representation (74) of metabolism was used in this study. Since only a small fraction of metabolites present in genome-scale metabolic models are typically measureable, a reduced model of metabolism was generated in which the fraction of the measured metabolites was enriched. A normalized Z-score for each reaction based on the Z-values of its neighboring metabolites was calculated, with the hypothesis that the Z-scores of the reactions calculated in this manner would indicate the significance of how a reaction responded to a perturbation at the metabolic level. This was based on the fact that metabolite levels are governed by changes in fluxes and the enzyme activities. Metabolome data from two different strains of *S. cerevisiae* in two different environmental conditions with glucose as the sole carbon source were considered. Using the reporter reaction analysis, it was possible to identify a number of reactions that were affected by the perturbations. The approach could distinguish between the effects of genetic perturbation in both the environmental conditions. It could also identify the results of the genetic changes around the genes that were perturbed.

The next step was to combine this information of reporter reactions with the array data. All reactions in the network were then classified as metabolically regulated, hierarchically regulated, a combination of the two, or unregulated. Metabolic regulation was described as regulation at the level of enzyme kinetics (i.e., changes of the metabolite levels) and hierarchical regulation denoted regulation of flux at the level of enzyme production or activity (i.e., transcription/translation/post-translational modifications). The transcript values of all genes encoding for the same reaction were summed, and the p-values of the transcripts were then calculated using a t-test with an unequal variance and further converted into Z-scores to enable comparison with the Z-scores of reactions based on metabolome data. The reactions where only the transcript Z-scores were changed significantly were considered to be points of possible hierarchical regulation, and reactions where only the metabolite-based Z-score was changed significantly were considered to be metabolically regulated. When both Z-scores were significant, regulation was shared at both the levels and when none of these scores was significant, it was inferred that the reaction was unregulated or unable to be determined.

Permutations of this method, all aimed at distinguishing enzymes exhibiting transcriptional or metabolic regulation, have been developed more recently using strictly model predictions of fluxes in the absence of metabolomics or fluxomics data. For example, Cakir et al. (75) use a weighted average of all elementary modes (pathways representing the edges of the feasible solution space (76)) to calculate the "control effective flux" for an organism growing under a given set of growth conditions. Bordel et al. (77) use a flux sampling method that randomly chooses 500 feasible flux distributions, and the mean value of each flux is used. Changes in these predicted fluxes between conditions are then compared with changes in gene expression, and the reactions characterized as transcriptionally regulated (correlation between flux and gene expression), post-transcriptionally regulated (changes in gene expression with no change in flux), or metabolically regulated (changes in flux but not in gene expression).

Focusing specifically on the relationship between enzymes and metabolites, Fendt and coworkers postulated different relationships depending on the concentration of metabolites participating in a reaction relative to the K_M of an enzyme catalyzing it (78). If the concentration is far below K_M, indicating excess enzyme and substrate limitation, then the substrate concentration should not vary with the amount of enzyme present. If the concentration is within one order of magnitude of K_M, as is predicted to be the case for most enzymes (79), then there could be a trade-off between metabolite and enzyme abundance resulting in a negative correlation between substrate and enzyme concentration. Finally, if the concentration of substrate is much higher than the K_M, the enzyme concentration is at a minimum for a given flux. In such a situation it would be expected that product metabolite concentrations exhibit a positive correlation with enzyme capacity. Clearly these rules are oversimplifications because they consider a single enzyme without context of other reactions that can affect metabolite concentration, but when looking across the measurable metabolome, patterns should emerge. To generate a variety of data to test these hypotheses, the authors conducted experiments with wild-type *S. cerevisiae* on glucose and a Gcr2p regulatory mutant that exhibits altered behavior of glycolysis and TCA cycle genes, enzymes, and metabolites (80,81). The fold changes in gene and protein levels for about 50 reactions were plotted against fold changes in concentrations of related metabolites, including substrates, products, and cofactors of these reactions. Significant negative correlations were observed for substrate metabolites and cofactors, supporting the hypothesis that both concentrations are near the K_M values of enzymes utilizing them. No significant correlation occurred with product metabolites or cofactors. To further test the generality of this hypothesis, the researchers grew the same strains on ethanol, a gluconeogenic substrate. The same correlations held, even though many of the substrate metabolites on glucose became product metabolites on ethanol, and vice versa. Finally, Fendt and coworkers created constructs to modulate four individual glycolytic steps using a Tet-repressed promoter. In all four cases, they found an increase in substrate concentration of at least 2-fold upon down-regulation of the corresponding enzyme by tetracycline addition. From this work the authors concluded that alterations in enzyme level are buffered by converse changes in substrate metabolite concentration, thus maintaining homeostasis in central metabolism (78). A similar response was observed in a yeast regulatory mutant that reduces the production of amino acid biosynthesis genes; lower amino acid concentrations resulted, restoring flux by the relief of allosteric inhibition (82).

Yizhak et al. developed a method termed integrative omics-metabolic analysis (IOMA) that uses a combination of enzyme levels and metabolite concentrations to derive constraints for FBA models (83). For a core set of reactions for which proteomic and complete metabolomic data are available, they used Michaelis–Menten-like rate equations to calculate enzyme saturation level (metabolomics) and enzyme relative V_{max} (proteomics). Given a baseline (wild-type) flux distribution, the kinetics then provided additional constraints to the FBA model, which was solved using quadratic programming (due to the nonlinearity of the kinetic equations). Using simulated omics data, the integrated model predicted flux distributions of *E. coli* deletion mutants better than FBA alone, although the method has yet to be tested with actual omics data.

There are many examples in the literature where microarrays were used to investigate the genetic basis of stress response in *E. coli* (84), including the response of *E. coli* to solvents and fuels such as butanol and isobutanol (85,86). Jozefczuk and coworkers take these microarray-based stress response studies a step further to determine how the metabolome responds to stress, and found that the metabolic response is similar in nature but surprisingly much more specific than the transcriptome response (87). Four different environmental perturbations were evaluated: oxidative stress, carbon source shift, heat stress, and cold shock. Concentrations of 95 compounds were determined at time points between 10 minutes and 4 hours post-shift, in parallel with microarray-based transcriptome data. In agreement with prior studies, the microarray data indicated a general slowdown of central metabolism and cell growth, and a tendency toward energy conservation. In agreement with these general trends, the concentrations of metabolites in glycolysis, the pentose phosphate pathway, and the TCA cycle decreased rapidly, accompanied by accumulation of most amino acids. Furthermore, the researchers found both transcriptomic and metabolic responses to be greatest in the first two time points after the shift. Since there would be a time lag before transcriptional changes are reflected in the metabolite concentrations, the coincidence of these two responses suggests that they are independent. Finally, there was more overlap among the different stresses observed in the transcriptional response than in the metabolic response, indicating that metabolic response is more specific. An interpretation is that metabolism has to react faster and in a more targeted way to prevent immediate damage from the stress, as opposed to the genetic response, which is more general and long term.

Jozefczuk and coworkers then used two statistical approaches to determine the level of coordination between the data sets: untargeted co-clustering, and a targeted method using prior biological knowledge with canonical correlation analysis (CCA). Applying the untargeted method to the entire data set, there was only about 10% similarity between metabolomic and transcriptomic clusters. Overrepresented in this data set were co-clustering between amino acids and genes encoding amino acid catabolic genes (Figure 3.8), whereas relatively few examples were found between amino acids and the corresponding biosynthetic genes. In the targeted approach, the researchers focused on glycolysis, the pentose phosphate pathway, the TCA cycle, respiration, and associated transcriptional regulators. Clear associations were found for the control condition (no stress), heat stress, and stationary phase, but not for the other conditions applied. The results are summarized in Figure 3.9. An unexpected association in the control case was the *mqo* gene (encoding malate-quinone oxidoreductase) with all TCA cycle intermediates and pyruvate, suggesting that the *mqo* gene product has a major function in regulating TCA cycle flux. Indeed, there is evidence that malate-quinone oxidoreductase, as opposed to malate dehydrogenase, is the major route of malate oxidation during optimal growth conditions (88). In the stationary phase, the association is lost and replaced by association of TCA metabolites with *frdCD* (fumarate reductase),

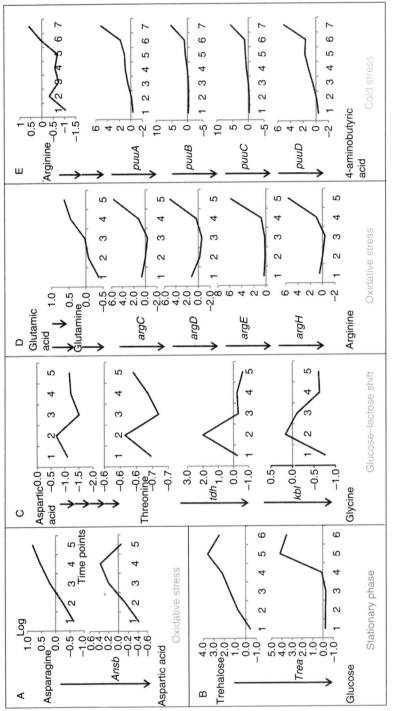

FIGURE 3.8. Co-clustering analysis revealed coordination between metabolite levels and transcriptional changes in amino acid catabolism pathways during the application of stress conditions. For each condition, exemplary pathways are drawn schematically to indicate key metabolites and genes. Changes in transcript levels are shown next to the genes, and changes in metabolite levels next to the metabolites, on a log scale. Time points correspond to every 10 minutes following application of the stress. Reprinted by permission from Macmillan Publishers Ltd.: *Molecular Systems Biology* (87), Copyright 2010.

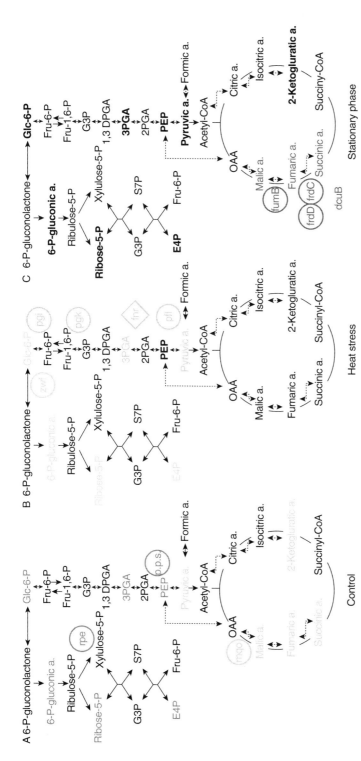

FIGURE 3.9. Canonical correlation analysis reveals condition-dependent association between response dynamics on the transcript and metabolite levels. Metabolites and genes showing close association were extracted and projected on a schematic representation of glycolysis, the pentose phosphate pathway, and the TCA cycle. Metabolic genes are circled and regulatory genes are displayed in a diamond box. Measured metabolites are indicated in bold. Transcripts and metabolites showing a close association are indicated by the same color. Reprinted by permission from Macmillan Publishers Ltd.: *Molecular Systems Biology* (87), Copyright 2010. (*See insert for color representation of the figure.*)

fumB (fumarase), and *dcuB* (fumarate-succinate antiporter). This suggests oxygen limitation and an induction of systems for alternate respiration (e.g., fumarate). Finally, the heat stress condition exhibits a strong association between pyruvate and genes involved in anaerobic fermentation (encoding pyruvate–formate lyase and the FNR transcriptional regulator), and with the glycolytic genes *glk* and *pgi* (87). Combined with earlier work on the *pgi* mutation (89), this suggests a complex role of these genes in anaerobic regulation. In a related publication, the same researchers looked at correlation among the different metabolites in stressed and non-stressed conditions (90). In each experiment a directed graph was constructed, with significant correlations depicted as edges between nodes (metabolites). The "stable network component" was defined as being the portions of these graphs common among all conditions. Szymanski et al. found a high degree of similarity of this component to the connectivity of these metabolites in the metabolic network (90). This finding suggests a possible application of metabolomics data in pathway reconstruction, particularly when combined with biochemical pathway prediction algorithms (91). Looking next at the correlations specific to each stress condition, they were able to identify key biomarkers as those metabolites that acted as central "hubs" (with many connections) in one condition only. An example is phosphoenolpyruvate (PEP), which in most conditions is an isolated node with little connectivity to other metabolites. Upon glucose starvation, however, PEP exhibits a high negative correlation with decreasing levels of malate, pyruvate, and amino acids. These observations suggest that PEP accumulation is a result of shifting metabolic resources from other parts of metabolism (90). Overall, application of clustering and correlation analysis on omics data is a systems-oriented approach to identifying the metabolic signals leading to complex regulatory mechanisms.

A final example is given by Lewis and coworkers (92), who used transcriptome and proteome analysis in conjunction with modeling to compare wild-type *E. coli* with strains evolved to grow optimally on different carbon sources (93). Using an approach termed parsimonious enzyme usage FBA (pFBA), they classified each gene based on its predicted contribution to the optimal growth phenotype in each growth condition. Genes are either essential for growth, essential for optimal growth, enzymatically or metabolically inefficient (contributing to lower growth prediction if used), or unable to carry flux at all. Omics data were then mapped onto the metabolic network and compared with the pFBA predictions. The essential and optimal genes were clearly overrepresented in both the proteins identified and the gene expression levels. Furthermore, expression levels exhibited a decreasing trend in the following order: essential > optimal > inefficient > no flux (92). The evolved strains had significantly increased growth rate and biomass yield on the carbon source used in the selection, which corresponded closely to the optimum growth point predicted by the constraint-based model (93). In this study it was also shown that this adaptation process repressed the pFBA-predicted no-flux genes (92).

3.8 FUTURE DIRECTIONS AND APPLICATIONS TO STRAIN ENGINEERING

The integration of high-throughput experimental data with metabolic modeling holds exciting prospects for the future in both discovery research and metabolic engineering. Omics techniques will improve in both quality and breadth of coverage, and will become more commonplace in the laboratory. Computational biologists will continue to develop methods such as those summarized here to interpret these high-quality data, particularly those aligned toward closing the gap between different levels of control. For example, as discussed earlier, it is clear that changes in gene expression do not always correlate with changes in metabolic flux. Models should be developed to include genetic regulation, translation, post-translational modification, and enzyme kinetics, thereby providing a theoretical framework for understanding this relationship. Finally, it will be important to develop methods that can reliably identify key points of flux control, as these would represent targets in metabolic engineering applications (94).

The principal critiques of constraint-based modeling have been the lack of regulatory information, and the restriction to steady-state. rFBA provides a framework for incorporating regulation in a Boolean sense, and for simulating the associated dynamics; however, it is restricted to known regulatory rules. Several methods are in development for the elucidation of regulatory rules from gene expression data (51,52,59), and could be used to develop rFBA networks for poorly characterized organisms. Further effort should also be spent developing integrated FBA models (53) to include kinetics of more regulatory and metabolic pathways. To date, the interest in such models has been rather limited primarily due to the lack of reliable kinetic data. Recently, rapid sampling techniques have been used to measure enzyme kinetics *in vivo* (95), and when not measureable kinetic constants can be extracted from fits of the model to experimental data. Metabolomics data can also be used for parameter identification (96).

One of the ultimate applications of constraint-based modeling is to drive strain development efforts for the production of pharmaceuticals, chemicals, and fuels (97). Several noteworthy examples of this are in the literature (98), but overall this breakthrough technology has been restricted to a small number of research groups. Classical strain improvement (CSI) by mutation and screening has been used to generate most of the production strains in commercial use today (see Chapters 1 and 6). Both rational engineering and CSI have a place in the next generation of strain development, and can often complement each other (99,100). The application of constraint-based modeling to fermentation data from existing production strains can drive subsequent rounds of rational engineering, and has been described (97). The addition of omics data, when applied judiciously at the right time and analyzed properly, has the potential of further accelerating development timelines. An example of how these technologies can be combined to engineer complexity is shown

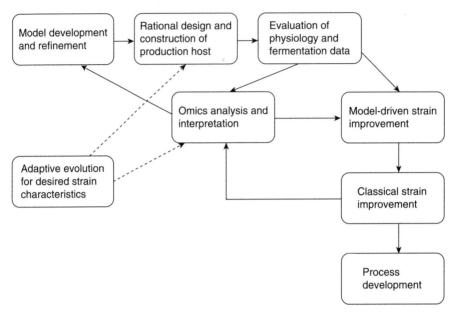

FIGURE 3.10. Use of constraint-based modeling and omics analysis to drive a combined rational and combinatorial strain development program. A first round of model-driven metabolic engineering is applied to develop an initial production strain. Fermentation products are measured and the model is used to guide subsequent improvements. High-throughput omics data can also be collected, analyzed by the methods discussed in this chapter, and used to identify additional targets. Constraints derived from these methods can also be used to refine the model. Once all rational manipulations have been made, classical strain improvement is applied. Analysis of the random modifications selected in this screen can lead to additional targets and further model refinement. The dotted lines show that the wild-type organism can optionally be improved by evolution prior to rational engineering.

in Figure 3.10. Evolution for desired strain properties (e.g., tolerance to product, substrate, impurities, or process conditions) can be performed in parallel to rational strain engineering. Once a strain has been engineered by model-driven directed manipulations, random mutagenesis and adaptive evolution can be applied to find improvement by means that could not have been predicted. Next-generation sequencing of the mutants will reveal sites of mutation, allowing hypotheses to be drawn about the function of mutated genes. In the case of mutated regulators, microarray analysis may suggest regulation targets. In a recent study of a classically generated production strain for clavulanic acid, microarray analysis showed significant increase of the expression of genes in central metabolism (99). More than half of the reactions predicted to correlate with increased clavulanic acid production by flux balance analysis had genes associated with them that were upregulated in the mutant. Furthermore, rationally engineered strains for clavulanic acid redirected flux in a similar way, but the classical mutant was actually closer to ideal because it allowed a small flux through glyceraldehyde-3-phosphate dehydrogenase, thus preventing the potential side effects of a complete deletion. Further rounds of rational engineering can then be performed based on these strains, and any novel information gained can be fed back into the model. Additional targets to achieve modified flux distributions can be identified using the techniques described in this chapter, applied to current production strains. Metabolomics data can find points of regulation, and in conjunction with gene expression data can be used to predict the type of regulation. As unnatural process conditions, such as high product concentration, can induce various stress responses in the cell, omics data can be used to elucidate the type of stress. Genetic targets will arise for deletion or overexpression, depending on whether the stress response should be enhanced or eliminated.

In conclusion, metabolic models can be utilized as scaffolds for the interpretation of omics data in the context of microbial strain development programs. Several methods for this analysis have been reported, and more are still in development. The best time to apply these methods is after the initial round of rational and/or classical strain development, so that the results of these manipulations can be observed and compared with the effects predicted by constraint-based modeling. Application of this approach will identify targets for further manipulations, which are often poorly characterized genes that would not have been identified using the models alone, and would be found much more quickly than with a strictly random approach.

REFERENCES

1. Palsson, B. (2002) *In silico* biology through "omics." *Nat Biotechnol*, **20**, 649–650.

2. Barrett, C.L., Kim, T.Y., Kim, H.U., Palsson, B.O., and Lee, S.Y. (2006) Systems biology as a foundation for genome-scale synthetic biology. *Curr Opin Biotechnol*, **17**, 488–492.

3. Palsson, B. (2000) The challenges of *in silico* biology. *Nat Biotechnol*, **18**, 1147–1150.

4. Bonarius, H.P.J., Schmid, G., and Tramper, J. (1997) Flux analysis of underdetermined metabolic networks: the quest for the missing constraints. *Trends Biotechnol*, **15**, 308–314.

5. Pramanik, J. and Keasling, J.D. (1997) Stoichiometric model of *Escherichia coli* metabolism: incorporation of growth-rate dependent biomass composition and mechanistic energy requirements. *Biotechnol Bioeng*, **56**, 398–421.

6. Varma, A. and Palsson, B.O. (1994) Stoichiometric flux balance models quantitatively predict growth and metabolic by-product secretion in wild-type *Escherichia coli* W3110. *Appl Environ Microbiol*, **60**, 3724–3731.

7. Covert, M.W., Schilling, C.H., Famili, I., Edwards, J.S., Goryanin, I.I., Selkov, E. et al. (2001) Metabolic modeling of microbial strains *in silico*. *Trends Biochem Sci*, **26**, 179–186.

8. Price, N.D., Reed, J.L., and Palsson, B.O. (2004) Genome-scale models of microbial cells: evaluating the consequences of constraints. *Nat Rev Microbiol*, **2**, 886–897.

9. Reed, J.L., Patel, T.R., Chen, K.H., Joyce, A.R., Applebee, M.K., Herring, C.D. et al. (2006) Systems approach to refining genome annotation. *Proc Natl Acad Sci U S A*, **103**, 17480–17484.

10. van der Werf, M.J., Overkamp, K.M., Muilwijk, B., Coulier, L., and Hankemeier, T. (2007) Microbial metabolomics: toward a platform with full metabolome coverage. *Anal Biochem*, **370**, 17–25.

11. Reed, J.L., Vo, T.D., Schilling, C.H., and Palsson, B.O. (2003) An expanded genome-scale model of *Escherichia coli* K-12 (iJR904 GSM/GPR). *Genome Biol*, **4**, R54.

12. Oh, Y.K., Palsson, B.O., Park, S.M., Schilling, C.H., and Mahadevan, R. (2007) Genome-scale reconstruction of metabolic network in Bacillus subtilis based on high-throughput phenotyping and gene essentiality data. *J Biol Chem*, **282**, 28791–28799.

13. Savinell, J.M. and Palsson, B.O. (1992) Network analysis of intermediary metabolism using linear optimization. I. Development of mathematical formalism. *J Theor Biol*, **154**, 421–454.

14. Klamt, S., Schuster, S., and Gilles, E.D. (2002) Calculability analysis in underdetermined metabolic networks illustrated by a model of the central metabolism in purple nonsulfur bacteria. *Biotechnol Bioeng*, **77**, 734–751.

15. Schilling, C.H., Schuster, S., Palsson, B.O., and Heinrich, R. (1999) Metabolic pathway analysis: basic concepts and scientific applications in the post-genomic era. *Biotechnol Prog*, **15**, 296–303.

16. Vallino, J.J. and Stephanopoulos, G. (1993) Metabolic fluc distributions in *Corynebacterium glutamicum* during growth and lysine overproduction. *Biotechnol Bioeng*, **41**, 633–646.

17. Kim, T.Y., Sohn, S.B., Kim, H.U., and Lee, S.Y. (2008) Strategies for systems-level metabolic engineering. *Biotechnol J*, **3**, 612–623.

18. Joyce, A.R. and Palsson, B.O. (2006) The model organism as a system: integrating "omics" data sets. *Nat Rev Mol Cell Biol*, **7**, 198–210.

19. DeRisi, J.L., Iyer, V.R., and Brown, P.O. (1997) Exploring the metabolic and genetic control of gene expression on a genomic scale. *Science*, **278**, 680–686.

20. Long, A.D., Mangalam, H.J., Chan, B.Y., Tolleri, L., Hatfield, G.W., and Baldi, P. (2001) Improved statistical inference from DNA microarray data using analysis of variance and a Bayesian statistical framework. Analysis of global gene expression in *Escherichia coli* K12. *J Biol Chem*, **276**, 19937–19944.

21. Troyanskaya, O.G., Garber, M.E., Brown, P.O., Botstein, D., and Altman, R.B. (2002) Nonparametric methods for identifying differentially expressed genes in microarray data. *Bioinformatics*, **18**, 1454–1461.

22. Sharov, A.A., Dudekula, D.B., and Ko, M.S. (2005) A web-based tool for principal component and significance analysis of microarray data. *Bioinformatics*, **21**, 2548–2549.

23. Wall, M.E., Dyck, P.A., and Brettin, T.S. (2001) SVDMAN—singular value decomposition analysis of microarray data. *Bioinformatics*, **17**, 566–568.

24. Tamayo, P., Slonim, D., Mesirov, J., Zhu, Q., Kitareewan, S., Dmitrovsky, E. et al. (1999) Interpreting patterns of gene expression with self-organizing maps: methods and application to hematopoietic differentiation. *Proc Natl Acad Sci U S A*, **96**, 2907–2912.

25. Eisen, M.B., Spellman, P.T., Brown, P.O., and Botstein, D. (1998) Cluster analysis and display of genome-wide expression patterns. *Proc Natl Acad Sci U S A*, **95**, 14863–14868.

26. Lockhart, D.J. and Winzeler, E.A. (2000) Genomics, gene expression and DNA arrays. *Nature*, **405**, 827–836.

27. Lewis, N.E., Cho, B.K., Knight, E.M., and Palsson, B.O. (2009) Gene expression profiling and the use of genome-scale *in silico* models of *Escherichia coli* for analysis: providing context for content. *J Bacteriol*, **191**, 3437–3444.

28. Bennett, B.D., Yuan, J., Kimball, E.H., and Rabinowitz, J.D. (2008) Absolute quantitation of intracellular metabolite concentrations by an isotope ratio-based approach. *Nat Protoc*, **3**, 1299–1311.

29. Lu, W., Kimball, E., and Rabinowitz, J.D. (2006) A high-performance liquid chromatography-tandem mass spectrometry method for quantitation of nitrogen-containing intracellular metabolites. *J Am Soc Mass Spectrom*, **17**, 37–50.

30. Lu, W., Clasquin, M.F., Melamud, E., Amador-Noguez, D., Caudy, A.A., and Rabinowitz, J.D. (2010) Metabolomic analysis via reversed-phase ion-pairing liquid chromatography coupled to a stand alone orbitrap mass spectrometer. *Anal Chem*, **82**, 3212–3221.

31. Jenkins, H., Hardy, N., Beckmann, M., Draper, J., Smith, A.R., Taylor, J. et al. (2004) A proposed framework for the description of plant metabolomics experiments and their results. *Nat Biotechnol*, **22**, 1601–1606.

32. Taylor, E.W., Jia, W., Bush, M., and Dollinger, G.D. (2002) Accelerating the drug optimization process: identification, structure elucidation, and quantification of *in vivo* metabolites using stable isotopes with LC/MSn and the chemiluminescent nitrogen detector. *Anal Chem*, **74**, 3232–3238.

33. van der Werf, M.J., Jellema, R.H., and Hankemeier, T. (2005) Microbial metabolomics: replacing trial-and-error by the unbiased selection and ranking of targets. *J Ind Microbiol Biotechnol*, **32**, 234–252.

34. van den Berg, R.A., Hoefsloot, H.C., Westerhuis, J.A., Smilde, A.K., and van der Werf, M.J. (2006) Centering, scaling, and transformations: improving the biological information content of metabolomics data. *BMC Genomics*, **7**, 142.

35. Van Deun, K., Smilde, A.K., van der Werf, M.J., Kiers, H.A., and Van Mechelen, I. (2009) A structured overview of simultaneous component based data integration. *BMC Bioinformatics*, **10**, 246.

36. Vaidyanathan, S., Broadhurst, D.I., Kell, D.B., and Goodacre, R. (2003) Explanatory optimization of protein mass spectrometry via genetic search. *Anal Chem*, **75**, 6679–6686.

37. Sweetlove, L.J., Last, R.L., and Fernie, A.R. (2003) Predictive metabolic engineering: a goal for systems biology. *Plant Physiol*, **132**, 420–425.

38. Van Dien, S. and Schilling, C.H. (2006) Bringing metabolomics data into the forefront of systems biology. *Mol Syst Biol*, **2**, 2006.

39. Evans, A.M., DeHaven, C.D., Barrett, T., Mitchell, M., and Milgram, E. (2009) Integrated, nontargeted ultrahigh performance liquid chromatography/ electrospray ionization tandem mass spectrometry platform for the identification and relative quantification of the small-molecule complement of biological systems. *Anal Chem*, **81**, 6656–6667.

40. Feist, A.M., Henry, C.S., Reed, J.L., Krummenacker, M., Joyce, A.R., Karp, P.D. et al. (2007) A genome-scale metabolic reconstruction for *Escherichia coli* K-12 MG1655 that accounts for 1260 ORFs and thermodynamic information. *Mol Syst Biol*, **3**, 121.

41. Henry, C.S., Jankowski, M.D., Broadbelt, L.J., and Hatzimanikatis, V. (2006) Genome-scale thermodynamic analysis of *Escherichia coli* metabolism. *Biophys J*, **90**, 1453–1461.

42. Mavrovouniotis, M.L. (1991) Estimation of standard Gibbs energy changes of biotransformations. *J Biol Chem*, **266**, 14440–14445.

43. Beard, D.A., Liang, S.C., and Qian, H. (2002) Energy balance for analysis of complex metabolic networks. *Biophys J*, **83**, 79–86.

44. Kummel, A., Panke, S., and Heinemann, M. (2006) Putative regulatory sites unraveled by network-embedded thermodynamic analysis of metabolome data. *Mol Syst Biol*, **2**, 2006.0034.

45. Henry, C.S., Broadbelt, L.J., and Hatzimanikatis, V. (2007) Thermodynamics-based metabolic flux analysis. *Biophys J*, **92**, 1792–1805.

46. Smith, M.W. and Neidhardt, F.C. (1983) Proteins induced by anaerobiosis in *Escherichia coli*. *J Bacteriol*, **154**, 336–343.

47. Epstein, W., Rothman-Denes, L.B., and Hesse, J. (1975) Adenosine 3′:5′-cyclic monophosphate as mediator of catabolite repression in *Escherichia coli*. *Proc Natl Acad Sci U S A*, **72**, 2300–2304.

48. Covert, M.W. and Palsson, B.O. (2002) Transcriptional regulation in constraints-based metabolic models of *Escherichia coli*. *J Biol Chem*, **277**, 28058–28064.

49. Covert, M.W., Knight, E.M., Reed, J.L., Herrgard, M.J., and Palsson, B.O. (2004) Integrating high-throughput and computational data elucidates bacterial networks. *Nature*, **429**, 92–96.

50. Herrgard, M.J., Lee, B.S., Portnoy, V., and Palsson, B.O. (2006) Integrated analysis of regulatory and metabolic networks reveals novel regulatory mechanisms in *Saccharomyces cerevisiae*. *Genome Res*, **16**, 627–635.

51. Reiss, D.J., Baliga, N.S., and Bonneau, R. (2006) Integrated biclustering of heterogeneous genome-wide datasets for the inference of global regulatory networks. *BMC Bioinformatics*, **7**, 280.

52. Bonneau, R., Reiss, D.J., Shannon, P., Facciotti, M., Hood, L., Baliga, N.S. et al. (2006) The Inferelator: an algorithm for learning parsimonious regulatory networks from systems-biology data sets de novo. *Genome Biol*, **7**, R36.

53. Covert, M.W., Xiao, N., Chen, T.J., and Karr, J.R. (2008) Integrating metabolic, transcriptional regulatory and signal transduction models in *Escherichia coli*. *Bioinformatics*, **24**, 2044–2050.

54. Lee, J.M., Gianchandani, E.P., Eddy, J.A., and Papin, J.A. (2008) Dynamic analysis of integrated signaling, metabolic, and regulatory networks. *PLoS Comput Biol*, **4**, e1000086.

55. Akesson, M., Forster, J., and Nielsen, J. (2004) Integration of gene expression data into genome-scale metabolic models. *Metab Eng*, **6**, 285–293.

56. Yang, C., Hua, Q., and Shimizu, K. (2002) Integration of the information from gene expression and metabolic fluxes for the analysis of the regulatory mechanisms in Synechocystis. *Appl Microbiol Biotechnol*, **58**, 813–822.

57. Tseng, C.P., Yu, C.C., Lin, H.H., Chang, C.Y., and Kuo, J.T. (2001) Oxygen- and growth rate-dependent regulation of *Escherichia coli* fumarase (FumA, FumB, and FumC) activity. *J Bacteriol*, **183**, 461–467.

58. Miller, E.N., Jarboe, L.R., Turner, P.C., Pharkya, P., Yomano, L.P., York, S.W. et al. (2009) Furfural inhibits growth by limiting sulfur assimilation in ethanologenic *Escherichia coli* strain LY180. *Appl Environ Microbiol*, **75**, 6132–6141.

59. Liao, J.C., Boscolo, R., Yang, Y.L., Tran, L.M., Sabatti, C., and Roychowdhury, V.P. (2003) Network component analysis: reconstruction of regulatory signals in biological systems. *Proc Natl Acad Sci U S A*, **100**, 15522–15527.

60. Jarboe, L.R. (2011) YqhD: a broad-substrate range aldehyde reductase with various applications in production of biorenewable fuels and chemicals. *Appl Microbiol Biotechnol*, **89**, 249–257.

61. Piper, M.D., Daran-Lapujade, P., Bro, C., Regenberg, B., Knudsen, S., Nielsen, J. et al. (2002) Reproducibility of oligonucleotide microarray transcriptome analyses. An interlaboratory comparison using chemostat cultures of *Saccharomyces cerevisiae*. *J Biol Chem*, **277**, 37001–37008.

62. Forster, J., Famili, I., Fu, P., Palsson, B.O., and Nielsen, J. (2003) Genome-scale reconstruction of the *Saccharomyces cerevisiae* metabolic network. *Genome Res*, **13**, 244–253.

63. Gombert, A.K., Moreira Dos, S.M., Christensen, B., and Nielsen, J. (2001) Network identification and flux quantification in the central metabolism of *Saccharomyces cerevisiae* under different conditions of glucose repression. *J Bacteriol*, **183**, 1441–1451.

64. Patil, K.R. and Nielsen, J. (2005) Uncovering transcriptional regulation of metabolism by using metabolic network topology. *Proc Natl Acad Sci U S A*, **102**, 2685–2689.

65. Hoppe, A., Hoffmann, S., and Holzhutter, H.G. (2007) Including metabolite concentrations into flux balance analysis: thermodynamic realizability as a constraint on flux distributions in metabolic networks. *BMC Syst Biol*, **1**, 23.

66. Jankowski, M.D., Henry, C.S., Broadbelt, L.J., and Hatzimanikatis, V. (2008) Group contribution method for thermodynamic analysis of complex metabolic networks. *Biophys J*, **95**, 1487–1499.

67. Visser, D., Schmid, J.W., Mauch, K., Reuss, M., and Heijnen, J.J. (2004) Optimal re-design of primary metabolism in *Escherichia coli* using linlog kinetics. *Metab Eng*, **6**, 378–390.

68. Nielsen, J. (1997) Metabolic control analysis of biochemical pathways based on a thermokinetic description of reaction rates. *Biochem J*, **321**(Pt 1), 133–138.

69. Wang, L., Birol, I., and Hatzimanikatis, V. (2004) Metabolic control analysis under uncertainty: framework development and case studies. *Biophys J*, **87**, 3750–3763.

70. Crabtree, B., Newsholme, E.A., and Reppas, N.B. (2010) Principles of regulation and control in biochemistry: a pragmatic, flux-oriented approach. In Hoffman, J.F. and Jamieson, J.D. (eds.), *Handbook of Physiology: Cell Physiology*. Oxford University Press, New York, pp. 117–180.

71. Kummel, A., Panke, S., and Heinemann, M. (2006) Systematic assignment of thermodynamic constraints in metabolic network models. *BMC Bioinformatics*, **7**, 512.

72. Bennett, B.D., Kimball, E.H., Gao, M., Osterhout, R., Van Dien, S.J., and Rabinowitz, J.D. (2009) Absolute metabolite concentrations and implied enzyme active site occupancy in *Escherichia coli*. *Nat Chem Biol*, **5**, 593–599.

73. Cakir, T., Patil, K.R., Onsan, Z., Ulgen, K.O., Kirdar, B., and Nielsen, J. (2006) Integration of metabolome data with metabolic networks reveals reporter reactions. *Mol Syst Biol*, **2**, 50.

74. Wagner, A. and Fell, D.A. (2001) The small world inside large metabolic networks. *Proc Biol Sci*, **268**, 1803–1810.

75. Cakir, T., Kirdar, B., Onsan, Z.I., Ulgen, K.O., and Nielsen, J. (2007) Effect of carbon source perturbations on transcriptional regulation of metabolic fluxes in *Saccharomyces cerevisiae*. *BMC Syst Biol*, **1**, 18.

76. Stelling, J., Klamt, S., Bettenbrock, K., Schuster, S., and Gilles, E.D. (2002) Metabolic network structure determines key aspects of functionality and regulation. *Nature*, **420**, 190–193.

77. Bordel, S., Agren, R., and Nielsen, J. (2010) Sampling the solution space in genome-scale metabolic networks reveals transcriptional regulation in key enzymes. *PLoS Comput Biol*, **6**, e1000859.

78. Fendt, S.M., Buescher, J.M., Rudroff, F., Picotti, P., Zamboni, N., and Sauer, U. (2010) Tradeoff between enzyme and metabolite efficiency maintains metabolic homeostasis upon perturbations in enzyme capacity. *Mol Syst Biol*, **6**, 356.

79. Cornish-Bowden, A. (1976) The effect of natural selection on enzymic catalysis. *J Mol Biol*, **101**, 1–9.

80. Uemura, H. and Fraenkel, D.G. (1990) gcr2, a new mutation affecting glycolytic gene expression in *Saccharomyces cerevisiae*. *Mol Cell Biol*, **10**, 6389–6396.

81. Uemura, H. and Fraenkel, D.G. (1999) Glucose metabolism in gcr mutants of *Saccharomyces cerevisiae*. *J Bacteriol*, **181**, 4719–4723.

82. Moxley, J.F., Jewett, M.C., Antoniewicz, M.R., Villas-Boas, S.G., Alper, H., Wheeler, R.T. et al. (2009) Linking high-resolution metabolic flux phenotypes and transcriptional regulation in yeast modulated by the global regulator Gcn4p. *Proc Natl Acad Sci U S A*, **106**, 6477–6482.

83. Yizhak, K., Benyamini, T., Liebermeister, W., Ruppin, E., and Shlomi, T. (2010) Integrating quantitative proteomics and metabolomics with a genome-scale metabolic network model. *Bioinformatics*, **26**, i255–i260.

84. Chang, D.E., Smalley, D.J., and Conway, T. (2002) Gene expression profiling of *Escherichia coli* growth transitions: an expanded stringent response model. *Mol Microbiol*, **45**, 289–306.

85. Rutherford, B.J., Dahl, R.H., Price, R.E., Szmidt, H.L., Benke, P.I., Mukhopadhyay, A. et al. (2010) Functional genomic study of exogenous n-butanol stress in *Escherichia coli. Appl Environ Microbiol*, **76**, 1935–1945.

86. Brynildsen, M.P. and Liao, J.C. (2009) An integrated network approach identifies the isobutanol response network of *Escherichia coli. Mol Syst Biol*, **5**, 277.

87. Jozefczuk, S., Klie, S., Catchpole, G., Szymanski, J., Cuadros-Inostroza, A., Steinhauser, D. et al. (2010) Metabolomic and transcriptomic stress response of *Escherichia coli. Mol Syst Biol*, **6**, 364.

88. van der Rest, M.E., Frank, C., and Molenaar, D. (2000) Functions of the membrane-associated and cytoplasmic malate dehydrogenases in the citric acid cycle of *Escherichia coli. J Bacteriol*, **182**, 6892–6899.

89. Rasmussen, L.J., Moller, P.L., and Atlung, T. (1991) Carbon metabolism regulates expression of the pfl (pyruvate formate-lyase) gene in *Escherichia coli. J Bacteriol*, **173**, 6390–6397.

90. Szymanski, J., Jozefczuk, S., Nikoloski, Z., Selbig, J., Nikiforova, V., Catchpole, G. et al. (2009) Stability of metabolic correlations under changing environmental conditions in *Escherichia coli*—a systems approach. *PLoS ONE*, **4**, e7441.

91. Hatzimanikatis, V., Li, C., Ionita, J.A., Henry, C.S., Jankowski, M.D., and Broadbelt, L.J. (2005) Exploring the diversity of complex metabolic networks. *Bioinformatics (Oxford, England)*, **21**, 1603–1609.

92. Lewis, N.E., Hixson, K.K., Conrad, T.M., Lerman, J.A., Charusanti, P., Polpitiya, A.D. et al. (2010) Omic data from evolved *E. coli* are consistent with computed optimal growth from genome-scale models. *Mol Syst Biol*, **6**, 390.

93. Fong, S.S., Joyce, A.R., and Palsson, B.O. (2005) Parallel adaptive evolution cultures of *Escherichia coli* lead to convergent growth phenotypes with different gene expression states. *Genome Res*, **15**, 1365–1372.

94. Reaves, M.L. and Rabinowitz, J.D. (2011) Metabolomics in systems microbiology. *Curr Opin Biotechnol*, **22**, 17–25.

95. Visser, D., van Zuylen, G.A., van Dam, J.C., Oudshoorn, A., Eman, M.R., Ras, C. et al. (2002) Rapid sampling for analysis of *in vivo* kinetics using the BioScope: a system for continuous-pulse experiments. *Biotechnol Bioeng*, **79**, 674–681.

96. Yuan, J., Doucette, C.D., Fowler, W.U., Feng, X.J., Piazza, M., Rabitz, H.A. et al. (2009) Metabolomics-driven quantitative analysis of ammonia assimilation in *E. coli. Mol Syst Biol*, **5**, 302.

97. Mahadevan, R., Burgard, A., Famili, I., Van Dien, S., and Schilling, C. (2005) Applications of metabolic modeling to drive bioprocess development for the production of value-added chemicals. *Biotechnol Bioprocess Eng*, **10**, 408–417.

98. Lee, S.Y., Lee, D.Y., and Kim, T.Y. (2005) Systems biotechnology for strain improvement. *Trends Biotechnol*, **23**, 349–358.

99. Medema, M.H., Alam, M.T., Heijne, W.H., van den Berg, M.A., Muller, U., Trefzer, A. et al. (2011) Genome-wide gene expression changes in an industrial clavulanic acid overproduction strain of Streptomyces clavuligerus. *Microb Biotechnol*, **4**, 300–305.

100. Alper, H., Miyaoku, K., and Stephanopoulos, G. (2005) Construction of lycopene-overproducing *E. coli* strains by combining systematic and combinatorial gene knockout targets. *Nat Biotechnol*, **23**, 612–616.

4

STRAIN IMPROVEMENT VIA EVOLUTIONARY ENGINEERING

Byoungjin Kim, Jing Du, and Huimin Zhao

4.0 INTRODUCTION

The development of industrially relevant microbial strains is a challenging task due to the complexity of microbial cells and of the phenotypes required for industrial processes. Rational approaches for strain improvement involve knockout or knock-in of specific target genes in the chromosome. However, complex phenotypes associated with multiple genes and their interactions are difficult to achieve by rational design targeting one or a few genes at a time (1). Although rapidly advancing genomic, proteomic, metabolic, and high-throughput analytical tools have significantly reduced the labor, time, and cost associated with strain engineering, the success of rational approaches largely depends on the detailed understanding of the biochemical and regulatory networks.

Evolutionary engineering mimics the natural evolutionary processes, consisting of iterative rounds of genetic diversification and functional selection or screening (Figure 4.1). Unlike rational design, evolutionary engineering is less dependent on prior knowledge of the phenotype–genotype relationship. Strain improvement is achieved by efficiently creating genetic diversity through mutagenesis (natural or induced) and recombination or shuffling of genes, pathways, and genomes, followed by high-throughput screening or selection for a desired phenotype. The strains used for industrial processes are often required to possess multiple phenotypes, such as tolerance to the metabolic

Engineering Complex Phenotypes in Industrial Strains, First Edition. Edited by Ranjan Patnaik.
© 2013 John Wiley & Sons, Inc. Published 2013 by John Wiley & Sons, Inc.

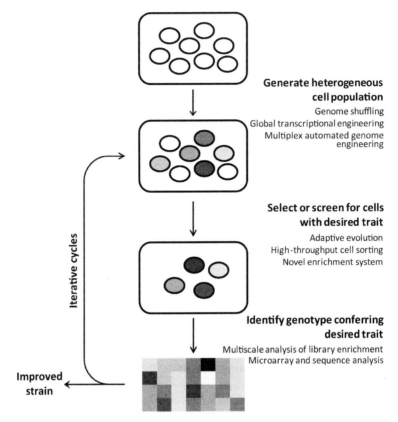

FIGURE 4.1. Scheme of the strain development process and the key methods available in each step.

product and inhibitors and high productivity, in order to meet the commercialization criteria. Evolutionary engineering can be performed for this purpose by evolving and identifying adapted strains using screening criteria that reflect feasible process conditions.

Creating genetic diversity covering a large sequence space and designing an efficient high-throughput screening or selection strategy are the two most critical steps in evolutionary engineering. Over the past decades, many advances have been made in the methodologies for the creation of genetically heterogeneous microbial populations, the automation of genome-wide sequence analysis, and high-throughput screening (1–5). In this chapter, some of these methodologies will be highlighted and a few representative examples will be discussed.

4.1 METHODOLOGIES FOR EVOLUTIONARY ENGINEERING

4.1.1 Adaptive Evolution

Adaptation refers to an evolutionary process allowing an organism's reproductive success in a given environment. In the adaptive evolution method, the evolutionary engineering cycle begins with the creation of a variant cell population, followed by selection or screening for desired phenotypes (Chapter 1). Generation of genetic diversity is achieved through the naturally occurring genetic variations in individual microorganisms and their continuous propagation through cellular replication. In this method, the evolutionary direction toward a desired phenotype is determined by a selection strategy. Adaptive evolution can be performed in batch or continuous cultures. In continuous cultures, variants with better fitness to a given environment outgrow over time and replace the parental population. In batch culture, a small fraction of the current culture (generally 10%) is transferred to fresh media before the depletion of nutrients, and this process is sequentially repeated until the targeted number of generations is reached. Cells in each batch culture pass through the lag, exponential, and stationary growth phases, and a significant change in growth (selection) conditions from nutrient-rich to nutrient-limited growth environments occurs in each cycle. In sequential batch cultures, it was found that significant fitness improvement occurred in the early stages of adaptive evolution experiments, and the rate of competitive fitness improvement hyperbolically decreased over time (6,7).

A chemostat is the most frequently used continuous culture system. In a chemostat, cells are in a physiological steady state. The cellular growth environment, including metabolite concentrations, growth rate, and cell density, is held constant, and it can be controlled by adjusting flow rates into the chemostat. These characteristics of a chemostat provide direct control of the selection pressure by modulating the culture conditions. Although the chemostat provides better control of the selection pressure than the batch culture system, care should be taken when the desired phenotype is not compatible with the improved growth rate (e.g., maximum growth vs. maximum production) because the continuous removal from the culture makes the growth rate a part of the selection function (8). The fitness can decrease as a result of epistatic interactions between multiple adaptive mutations in continuous asexual cultures and consequently it is possible that certain adaptive mutations may repeatedly appear and disappear, limiting the repertoire of adaptive mutations (9). To avoid this issue and ensure fitness improvement over time, it is necessary to monitor evolutionary progress by characterizing the phenotypes of the variants throughout the culture (8). In addition to the above discussed classical approaches for creating genetic diversity, newer approaches for creating diversity in microbial populations, as discussed in Sections 4.1.2 to 4.1.8, have rendered the engineering of multigenic complex phenotype via evolutionary approaches more systematic and predictable.

4.1.2 Genome Shuffling

Genome shuffling is achieved by homologous recombination between genomes through protoplast fusion and therefore multiple genes across the entire genome can be modified simultaneously (10,11). Since the 1970s, protoplast fusion has been used for strain improvement and was shown to be applicable to both prokaryotic and eukaryotic cells to obtain a high frequency of gene transfer (12–15). In protoplast fusion, protoplasts are isolated from cells by digesting the cell wall in the presence of osmotic stabilizers, whereas the fusion of protoplasts is induced by fusogen such as polyethylene glycol (Figure 4.2a). Unlike classical breeding, which enables recombination between only two parents per generation, genome shuffling can be carried out between multiple parents, and recursive shuffling can further accelerate the evolution process, producing multi-parent complex progeny (Figure 4.2b) (10). Since genome shuffling exploits the diversity and sequence homology given by the parental strains, the presence of highly homologous regions on the parental genomes can cause biased recombination producing representative progenies. This issue can be alleviated by taking combinatorial approaches with classical strain improvement strategies such as random mutagenesis and chemostat enrichments. After isolating a desired phenotype by screening and selection,

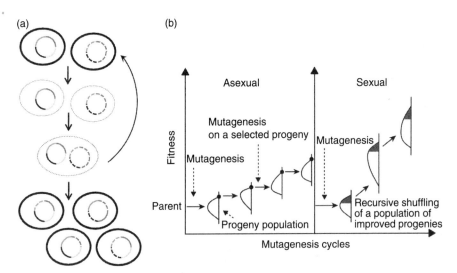

FIGURE 4.2. Schematic diagram of genome shuffling showing (a) cell-wall digestion, protoplast fusion, and the resulting heterogeneous cell population, and (b) fitness improvement in asexual and sexual evolution.

accumulated nonessential or deleterious mutations during rounds of random mutagenesis can be removed by back-crossing of progeny to parents (3). Genome shuffling can be particularly useful for industrial strain development where ill-characterized strains need to be engineered for complex phenotypes in a constrained time frame (16). Additional formats for genome shuffling as applied to engineering of tolerance in *Saccharomyces cerevisiae* to inhibitors is discussed in Chapter 9.

4.1.3 Global Transcriptional Machinery Engineering

Global transcriptional machinery engineering (gTME) alters the proteins regulating the global transcriptome and generates diversity at the transcriptional level, producing a pool of variants with heterogeneous phenotypes (Figure 4.3). Instead of modifying genes, gTME aims at perturbing the expression of multiple proteins simultaneously by creating a mutant library of the protein coordinating them. The sigma factor (σ^{70}) in *E. coli* and the RNA Pol II transcription factor D (TFIID) component Spt15 in yeast were engineered by gTME to improve product tolerance, metabolite overproduction, and xylose utilization (5,17–19). Mutant libraries of global transcriptional machineries can be generated using traditional mutagenesis techniques such as error-prone PCR (20,21), and strains with the desired phenotype are obtained by subsequent screening or selection. Since gTME perturbs the expression of

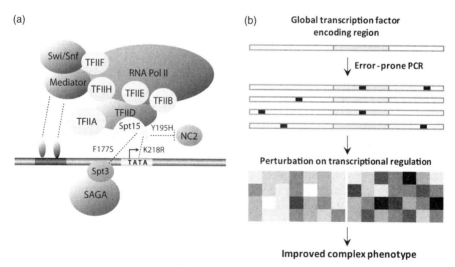

FIGURE 4.3. Global transcription machinery engineering (gTME). (a) From Alper and coworkers (17). Reprinted with permission from AAAS. Schematic diagram of global transcription machinery in *Saccharomyces cerevisiae*. (b) Key steps in gTME. The three mutations responsible for improved ethanol tolerance and production and proposed global transcription mechanism are shown in panel (a).

multiple genes simultaneously, it is efficient to exploit the mutation space for complex phenotypes involving multiple gene modifications across the genome. In addition, gTME combined with microarray-based transcriptional analysis can provide insight into the genotype–phenotype correlation, which is valuable for further strain improvement (17,18).

4.1.4   Transposon Insertion Mutagenesis

Gene deletion analysis has been used and proven to be an essential technique in determining the functionalities of genes (22,23). Distinct from the traditional gene deletion strategy, transposon insertion mutagenesis utilizes a mobile gene element, the transposon, to create a mutation library on the genome scale. The transposon is used as an insertional mutagen to disrupt gene activity and also serves as a tag that can be easily detected for the identification of the mutated gene (24). Commonly used transposon insertion mutagenesis systems share bacteria-derived transposons (e.g., Tn3, Tn5, and Tn7), antibiotic or auxotrophic markers for selection, and a reporter (e.g., *lacZ*). These gene elements are constructed into a plasmid containing a gene encoding transposase. A transposon mutant library can be generated *in vivo* or *in vitro* in bacterial microorganisms (25–27) and by shuttle mutagenesis for *S. cerevisiae* (24,28). In the latter, transpositions are generated in a library of yeast genomic DNA, and the mutated alleles are shuttled into yeast for subsequent analysis (29). The resulting library is screened under various growth conditions for the desired phenotype, and positive mutants are further analyzed to identify the genes responsible for the phenotypes using the transposon sequence as a tag sequence (27,30). Transposon mutagenesis was used to identify genes involved in cell viability, auxotrophy, antibiotic sensitivity, mobility, and growth on various growth conditions in organisms including *Mycobacteria*, *Escherichia coli*, *Pseudomonas aeruginosa*, and *S. cerevisiae* (24,26–28,30). Transposon mutagenesis was also used to identify two loci, *PHO13* and a region 500 bp upstream from the *TAL1* ORF, as responsible for the improved xylose utilization and fermentation by a recombinant *S. cerivisiae* strain (31). Transposon insertion mutagenesis shares its advantages with gTME and genome shuffling for the development of industrially relevant strains with complex phenotypes. Similar to gTME and genome shuffling, transposon mutagenesis combined with rapidly advancing microarray technologies will provide an efficient way of identifying the genotype–phenotype relationship on the genome scale.

4.1.5   Multiplex Automated Genome Engineering

Multiplex automated genome engineering (MAGE) offers an efficient method to create genetic diversity utilizing oligomer synthesis technology and automated cell culture and transformation (32). In one format of MAGE, synthetic DNAs that consist of degenerate nucleotide oligomer pools targeting specific

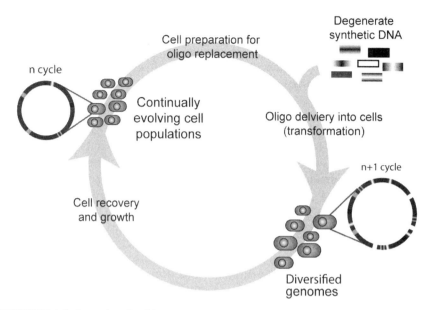

FIGURE 4.4. Steps involved in multiplex automated genome engineering (MAGE).

genomic positions are repeatedly introduced into a continuously evolving cell population, and the genetic diversity increases as the number of cycle increases (Figure 4.4). Through each cycle, synthetic DNA can be introduced into >30% of the cell population every 2–2.5 hours (32) and populations covering a large area of sequence space can be readily generated through successive MAGE cycles. Oligo-mediated allele replacements can produce mismatch, insertion, and deletion mutations on the genome, and the efficiency of allele replacement is determined by the type of mutation and the scale of the genetic modification. Using MAGE, multiple locations on the chromosome in a single cell can be targeted simultaneously across a population of cells. Because the genetic target and type of changes can be controlled by the design of the synthetic DNAs, MAGE can be flexibly applied to evaluating the effects of a single gene modification or a combination of multiple gene modifications on the fitness of the evolved strains. MAGE requires the sequence information of target genes, and when target alleles are properly selected, this method can produce a desirable phenotype on a shorter time scale compared with the rational metabolic engineering approach.

4.1.6 Tractable Multiplex Recombineering

Tractable multiplex recombineering (TRMR) aims at obtaining a comprehensive map of genetic modifications affecting a trait of interest by simultaneous creation and evaluation of specific genetic modifications on a genome scale (33). TRMR utilizes homologous recombination (recombineering) of barcoded oligonucleotides to modify a genetic network at the gene expression level. The barcoded oligonucleotides (synDNA) feature targeting oligos and functional cassettes. Targeting oligos contain homology regions and barcodes specific to each target gene and can include gene expression modulators. Functional cassettes are the sequences inserted upstream of the genes replacing the translation start codon and designed to modify gene expressions. A mixture of synDNA is transformed into a target organism. Upon successful recombineering, gene expression levels are controlled by the functional cassettes inserted. Screening or selection on a mixture of variants is followed to isolate a variant possessing a desired trait. Using barcode technology based on microarray (34), the frequency changes of the alleles during the enrichment or screening can be monitored and used to map a specific genetic change on the desired trait(s). It was demonstrated that every protein-coding gene in *E. coli* (4077 genes) could be targeted and successfully modified in the expression by using a conventional transformation technique. Currently, oligomer synthesis and transformation efficiency of various host organisms are the major challenges of this method. As the multiplex DNA synthesis and associated molecular biology techniques are further improved, TRMR might be broadly applicable in metabolic engineering and systems biology by providing a fast and efficient way of identifying a unique or a set of genetic modifications leading to a trait of interest. TRMR can be performed without *a priori* knowledge of the functions of the target genes and can be combined with MAGE or directed evolution to engineer microorganisms of complex traits.

4.1.7 Chemically Induced Chromosomal Evolution

Chemically induced chromosomal evolution (CIChE) provides a useful technique for maintaining a heterologous pathway on the chromosome to produce desired chemicals (35). Stable expression of a gene construct is critical to minimizing the fluctuations of productivity and product yield in industrial processes. CIChE is carried out by inducing chromosome evolution to contain multiple copy numbers of a gene cassette using increasing concentrations of antibiotics. Chromosome evolution is accomplished by propagating the CIChE cassette of the target genes on the chromosome by *recA*-mediated DNA crossover (35). The CIChE cassette contains the genes of interest, a selectable marker, and flanking homologous regions so that the cassette is propagated in a tandem manner. A desired copy number on the chromosome can be achieved by increasing the selection pressure (e.g., the antibiotic concentration). After the target number of generations is reached, *recA* can be deleted, and no further selection pressure is required to maintain the recombinant alleles (35), which can be a major advantage for the production on an industrial scale.

Genetic stability is another advantage of CIChE. In a long-term subculturing, test of strains producing the metabolically demanding product poly-2-hydroxybutyrate (PHB), the productivity of the plasmid-carrying strain dropped to zero after 40 generations, while the CIChE-engineered strain maintained the productivity (>90%) during the entire test (70 generations) (35). CIChE can be applicable for most industrially relevant host organisms in which the methods for genomic integration of the CIChE cassette, recombination knockout, and a *recA* homolog that can turn homologous recombination on and off are available.

4.1.8 Multiscale Analysis of Library Enrichment (SCALE)

Advances in technologies allowing fast and efficient generation of genetic diversity should be accompanied by technologies available for the identification of the genetic basis of the phenotype. Screening of a genomic DNA library under selective conditions involves creation of a genomic DNA library and identification of the genomic DNA fragments responsible for the phenotypes. Multiple subcloning steps are often required to isolate all genetic factors and their combinatorial roles in expression of the desired phenotype. Because of its laborious and time-consuming nature, it can delay the time frame for industrial strain development. The advantage of SCALE is its capability to identify multiple genes, which can be single short DNA sequences or operons, in a single experiment, eliminating time-consuming subcloning steps. Multiple populations representing genomic libraries of differently sized DNA fragments in plasmids are generated separately and mixed, resulting in a cell population containing multiple inserts of different lengths (36,37). Continuous selection is performed on the mixed library and the enriched gene (short insert) or operon (long insert) is identified by microarray analysis. The generation of a mixed library of genomic DNA inserts varying in length (500–8000 base pairs) and the capability of deconvoluting the microarray signal contributions from each of the different clones allow the identification of the locations as well as the sizes of the relevant DNA fragments responsible for the altered phenotype. The ability to obtain a truly representative genomic DNA libraries and the reliability of microarray analysis are the two most critical factors for successful implementation of this method. Application of SCALE technology for engineering tolerance to 3-hydroxypropionic acid in *E. coli* is discussed in Chapter 7.

4.1.9 Screening and Selection

As discussed in Chapter 1, efficient screening and selection techniques are critical to the isolation of variant strains with the desired phenotypes within the population. Traditional screening and selection involves isolation and analysis of individual variant strains for the desired phenotype. The availability

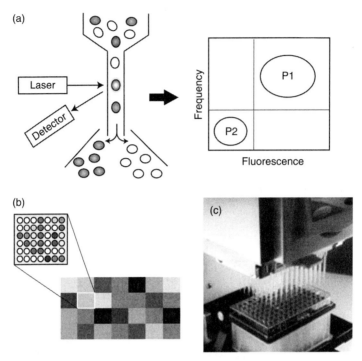

FIGURE 4.5. Technologies applicable for high-throughput analysis: (a) fluorescence-activated cell sorting, (b) genome-scale microarray, and (c) automatic sample handling device (Copyright © CyBio AG, Jena, Germany).

of assays and sensitive analytical techniques determine the efficiency of the screening and selection process and the resolving power to isolate a variant strain with the best fitness in a given condition (38). Rapidly advancing "omics" technologies and automation in sample handling and analysis have increased the throughput of classical screening and selection tools (Figure 4.5). Fluorescence-activated cell sorting (FACS) enables high-throughput screening of a large library at a single cell level, and this technique is particularly useful when the phenotype can be coupled with fluorescence or unique light scattering patterns (39,40) (Figure 4.5a). Pre-screening and selection can be used to reduce the final size of the library to be tested and to increase the chance to isolate a variant strain with the target phenotype with reduced cost and labor (38,41). In adaptive evolution, it is important to monitor the evolving population to ensure the evolutionary direction toward the global optimum. The time and cost for genome sequencing and microarray technology have decreased significantly over the past decade, and these technologies are anticipated to become even more widely accessible in the near future. Screening and selection tools combined with these new technologies would increase the efficiency of evolutionary engineering approaches for the development of industrial strains (Figure 4.5b,c).

4.2 EXAMPLES OF EVOLUTIONARY ENGINEERING

Evolutionary engineering approaches have been proven to be effective in creating industrial microorganisms with improved complex phenotypes, such as enhancement of product yield and productivity, extension of substrate range, and improvement of cellular properties (3,8). In this part of the chapter, recent progress in using evolutionary engineering approaches for strain improvement will be discussed. Development of high-throughput screening methods and their potential application for strain engineering will also be briefly mentioned.

4.2.1 Enhancement of Product Yield and Productivity

Mostly driven by environmental and energy security considerations, there is a growing interest in developing biocatalytic approaches for production of fuels and chemicals from renewable feedstock. However, only a few biotechnology based approaches have proven economically feasible for production of chemicals (42,43). The efficient production of chemicals in microorganisms can be limited by complex metabolic pathways, substrate and intermediate inhibition, and other fermentation by-products (44).

One successful example of using evolutionary engineering to improve product yield and productivity in microorganisms is the biochemical production of 1,3-propanediol (1,3-PD). 1,3-PD is a useful monomer for production of several plastics including polytrimethylene terephthalate (45). Otte and coworkers started with *Clostridium diolis* DSM 15410, a microorganism with good molar yield and volumetric productivity for conversion of glycerol into 1,3-PD under anaerobic conditions, and applied the genome shuffling method to optimize the substrate and product tolerance and the 1,3-PD productivity. A mutant library was generated using chemical mutagenesis. Mutants with higher substrate and product tolerance and higher product yields were isolated and used as parental strains for genome shuffling. Significant improvements in 1,3-PD productivity were observed after four rounds of genome shuffling and selection. The best mutant exhibited an 80% improvement in yield compared with the parental wild-type strain, and the final titer of 1,3-PD reached 85 g/L (46).

Another successful evolutionary engineering approach for the overproduction of valuable compounds is the gTME method. Although genetic regulation has been studied for the engineering of metabolic pathways, no work has been carried out to engineer metabolic pathways through the manipulation of global regulatory pathways until recently. Tatarko and coworkers engineered a carbon storage regulator (Csr), a global regulatory system of *E. coli*, to improve phenylalanine biosynthesis. The engineered strain with an optimized aromatic pathway produced twofold more phenylalanine when *csrA* was disrupted. This work is the first known example of metabolic engineering utilizing a global regulatory pathway, and it also introduced the concept of "global metabolic engineering" for the first time (47). Alper and coworkers randomly mutagenized the *rpoD* gene that encodes the main sigma factor σ^{70} in *E. coli*. The resulting library of *rpoD* variants was introduced into lycopene-producing *E. coli* with different gene deletion backgrounds. After a single round of gTME, several mutants with increased lycopene productivity were identified. The best mutants from *E. coli* with different gene deletion backgrounds harbored different mutated versions of *rpoD*. The lycopene content of several mutant strains after a 15-hour fermentation achieved similar increases compared with previously engineered multiple gene knockout strains (Figure 4.6).

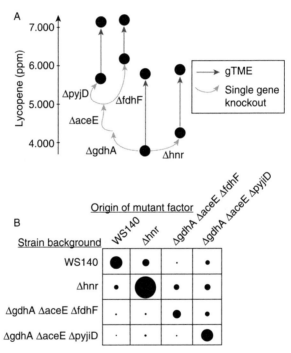

FIGURE 4.6. Application of gTME to a metabolite production phenotype. (A) Lycopene content, in mg/g dry cell weight (ppm) after 15-hour cultivations. The center of the black dots represents the production level of lycopene in ppm for a given strain, with the wild-type strain labeled at the bottom of the graph. The arrowheads of curved arrows not terminating at a black dot (e.g., gdhA knockout curve) represent the lycopene production of this strain. (B) A dot plot for each of the 16 strains is shown, which depicts the maximum fold increase achieved in lycopene production. The size of the circle is proportional to the fold increase. Reprinted from *Metabolic Engineering* **9**(3), 258–267, Copyright 2007, with permission from Elsevier.

It was shown that a single round of selection using gTME is more effective than multiple rounds of single-gene knockout or overexpression (18).

The MAGE method has also been used to improve the production of lycopene through optimization of the 1-deoxy-D-xylulose-5-phosphate (DXP) biosynthesis pathway in *E. coli*. Using synthetic DNA, 24 genetic components in the DXP pathway were modified simultaneously. The researchers constructed prototype devices that automate the MAGE technology to facilitate the rapid and continuous generation of a set of genetic changes including mismatches, insertions, and deletions. Using a complex pool of synthetic DNA, facilitated by the automated devices, the mutant library was generated over 4.3 billion combinatorial genomic variants per day. Mutants that showed more than a fivefold increase in lycopene production within 3 days of fermentation were isolated (32).

4.2.2 Extension of Substrate Range

In recent years, much effort has been spent on the engineering of microorganisms to convert lignocellulosic biomass into fuels (2). One of the microorganisms under intensive investigation is *S. cerevisiae*, as it is the microorganism currently used for large-scale ethanol production. As a eukaryotic organism, the metabolic network of *S. cerevisiae* is very complicated. In order to engineer *S. cerevisiae* to utilize lignocellulosic biomass efficiently, heterologous pathways have been introduced to enable assimilation of five-carbon sugars such as D-xylose and L-arabinose (48,49). However, the sugar utilization and ethanol production of recombinant yeast strains are still not efficient enough, and multiple properties of industrial yeast strains have to be modified. Evolutionary engineering approaches have been applied to improve the productivity of recombinant yeast strains for ethanol production by extending the substrate range from glucose to other sugars (50).

To improve the fermentation of glucose, D-xylose and L-arabinose mixtures by engineered *S. cerevisiae* strains, Wisselink and coworkers applied a novel evolutionary engineering approach involving repeated batch cultivation with repeated cycles of consecutive growth in media consisting of different sugar compositions (49). The strains were evolved in a mixture of sugars containing first glucose, D-xylose, and L-arabinose, then D-xylose and L-arabinose, and lastly L-arabinose only. The evolved strains can completely ferment a mixture of sugars containing 30 g/L glucose, 15 g/L D-xylose, and 15 g/L L-arabinose in 40% less time (49).

Kuyper and coworkers demonstrated that by introducing a heterologous xylose isomerase into *S. cerevisiae*, the resulting recombinant yeast strain can grow on D-xylose without the redox imbalance issue of the fungal D-xylose utilizing pathway (51). After a prolonged cultivation on D-xylose, a mutant strain that grew aerobically and anaerobically on D-xylose was obtained. The anaerobic ethanol yield reached 0.42 g ethanol per gram of D-xylose, and the by-product formation was also at a comparable level with the glucose-grown anaerobic culture. This study demonstrated that by using an evolutionary engineering approach, the enzyme activities and/or regulatory properties of native *S. cerevisiae* gene products can be optimized for D-xylose utilization under anaerobic conditions (51).

Sonderegger and Sauer started with the recombinant *S. cerevisiae* strain TMB3001 that overexpresses the fungal D-xylose utilization pathway from *Pichia stipitis*, and obtained an evolved strain that can grow anaerobically on D-xylose (52). They first selected organisms for efficient aerobic growth on D-xylose alone and then slowly adapted the organisms to microaerobic conditions, and finally anaerobic conditions. After a total of 460 generations or 266 days of selection, the culture consisted of primarily two subpopulations with distinct phenotypes that can be reproduced stably under anaerobic conditions on D-xylose. Further analysis of the two subpopulations revealed that only the larger subpopulation can grow anaerobically on D-xylose, while the smaller subpopulation, which was incapable of anaerobic growth, exhibited an improved D-xylose catabolism (52).

The construction of *S. cerevisiae* strains that ferment lactose is also useful for cheese whey fermentation. A recombinant flocculent lactose-consuming *S. cerevisiae* strain expressing the *LAC12* (lactose permease) and *LAC4* (β-galactosidase) genes of *Kluyveromyces lactis* has been constructed, but the lactose fermentation efficiency is suboptimal. Guimaraes and coworkers applied an evolutionary engineering process, that is, serial transfer and dilution in lactose medium, and yielded an evolved recombinant strain (53). The evolved strain can consume lactose twofold faster, and produce 30% more ethanol than the original recombinant. The researchers then investigated the evolved strain and identified two molecular events that targeted the *LAC* construct: a 1593-bp deletion in the promoter region between *LAC4* and *LAC12* and a decrease in the plasmid copy number by about 10-fold compared with the parental strain. The results suggest that the evolved promoter enabled the transcription of *LAC4* and *LAC12*. Together with the decreased copy number of both genes, the different levels of transcriptional induction for *LAC4* and *LAC12* improved lactose utilization in the evolved strain. The evolved strain obtained by simple adaptive engineering can efficiently ferment threefold more concentrated cheese whey, and provided an attractive alternative to the fermentation of lactose-based media (53).

4.2.3 Improvement of Cellular Properties

For industrial microorganisms, resistance to stresses is highly desirable due to the simultaneous or sequential combinations of different environmental stresses present in biotechnological processes. The molecular basis of stress resistances is complicated, making it difficult to engineer stress resistance by rational approaches (50). However, using evolutionary engineering approaches, engineering strains with multiple-stress resistances is possible. Cakar and coworkers tested various selection procedures in chemostats and

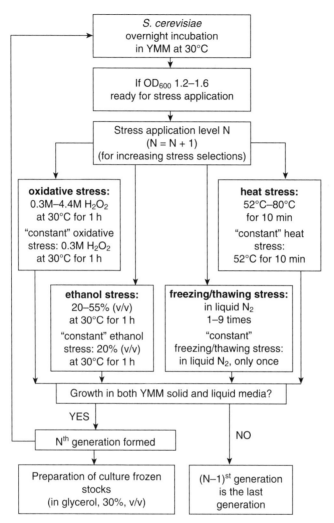

FIGURE 4.7. The algorithm for batch stress applications and formation of stress-resistant mutants. The term "constant" indicates constant, mild stress selection strategy for that particular stress (oxidative, ethanol, heat or freezing–thawing) resistance. YMM, yeast minimal medium. Reprinted from *FEMS Yeast Research*, **5**(6–7), 569–578, Copyright 2005, with permission from John Wiley & Sons, Inc.

batch cultures systematically for a multiple-stress resistance phenotype in *S. cerevisiae* (Figure 4.7) (50). Mutant populations were harvested at different time points and clones were randomly chosen and grown in batch cultures that were exposed to oxidative, freezing–thawing, high-temperature, and ethanol stress. A special procedure involving the use of a 96-well plate-based high-throughput screening method combined with a most-probable-number assay was developed for the selection of multi-stress resistant strains. In this study, the best selection strategy to obtain highly improved multiple-stress-resistant yeast was found to be batch selection for the freezing–thawing stress. Although mutants were selected toward freezing–thawing stress in this strategy, the resulting mutants were significantly improved not only in freezing–thawing stress resistance, but also in the other stress resistances mentioned earlier. The best isolated strain exhibited 102-, 89-, 62-, and 1429-fold increased resistance to freezing–thawing, temperature, ethanol, and oxidative stress, respectively (50).

Genome shuffling is another method that has been applied to improve microbial cellular properties such as ethanol tolerance (54), acetic acid tolerance (55), and tolerance to other inhibitors (56). Bajwa and coworkers (56) applied genome shuffling based on cross mating to improve the tolerance of fermentation ability of *Pichia stipitis* toward hardwood spent sulphite (HW SSL) (see also Chapter 9). After four rounds of genome shuffling, the mutants were able to produce ethanol from xylose present in undiluted HW SSL (56). The genome shuffling method has been used to improve the tolerance of microorganisms toward pesticides and the degradation of pesticides using *Sphingobium chlorophenolicum*. Pentachlorophenol (PCP) is a highly toxic anthropogenic pesticide, which can be mineralized and degraded by the gram-negative bacterium *S. chlorophenolicum*. However, wild-type *S. chlorophenolicum* can only degrade PCP at a very slow rate because of low tolerance to the compound. Dai and coworkers generated a mutant library using genome shuffling, and mutants with higher PCP tolerance were identified. After three rounds of genome shuffling using protoplast fusion, the mutant strains exhibited more than a 10-fold increase in tolerance toward PCP. Some mutant strains can also degrade 3 mM PCP in one-quarter-strength tryptic soy broth, while no degradation can be observed by their parental strains under the same condition (57).

4.3   CONCLUSIONS AND FUTURE PROSPECTS

Microorganisms have become increasingly exploited to address some of the most challenging global problems such as global warming, energy security, severe pollutions, and environmental degradation (4,58). In many cases, microorganisms used for industrial applications require multiple complex phenotypes such as high tolerance to stresses, substrates, products, pH, and temperatures. Evolutionary engineering approaches offer a promising alternative to traditional strain improvement methodologies in coping with this challenge. By harnessing the natural algorithm to select the fittest variants through continuous evolution, evolutionary engineering allows a more efficient and comprehensive searches across the rugged fitness landscape. The evolutionary pathway can be modulated depending on the type and complexity of the phenotype by carefully designing the screening and selection strategy over the course of evolution, which will enable the development and isolation of a phenotype compatible with the industrial operation conditions and facilitate scale-up from bench-top to production line. In addition, because the genetic backgrounds and phenotypic characteristics of many industrial strains are not available for rational metabolic engineering, evolutionary engineering will continue to be a valuable metabolic engineering strategy. By complementing rational approaches and emerging genome and transcriptome analysis tools, evolutionary engineering approaches will also expand our understanding of the genotype–phenotype relationship, providing new insights for further strain engineering.

ACKNOWLEDGMENTS

We thank the National Institutes of Health (GM077596), the National Academies Keck Futures Initiative on Synthetic Biology, the Biotechnology Research and Development Consortium (BRDC) (Project 2-4-121), the British Petroleum Energy Biosciences Institute, and the National Science Foundation for financial support in our synthetic biology projects. J. Du also acknowledges support from Chia-Chen Chu Graduate Fellowship in the School of Chemical Sciences at the University of Illinois at Urbana-Champaign.

REFERENCES

1. Warner, J.R., Patnaik, R., and Gill, R.T. (2009) Genomics enabled approaches in strain engineering. *Curr Opin Microbiol*, **12**, 223–230.
2. Dale, B.E. (2003) "Greening" the chemical industry: research and development priorities for biobased industrial products. *J Chem Technol Biotechnol*, **78**, 1093–1103.
3. Patnaik, R. (2008) Engineering complex phenotypes in industrial strains. *Biotechnol Prog*, **24**, 38–47.
4. Tang, W. and Zhao, H. (2009) Industrial biotechnology: tools and applications. *Biotechnol J*, **4**, 1725–1739.
5. Tyo, K.E., Alper, H.S., and Stephanopoulos, G.N. (2007) Expanding the metabolic engineering toolbox: more options to engineer cells. *Trends Biotechnol*, **25**, 132–137.
6. Lenski, R.E., Mongold, J.A., Sniegowski, P.D., Travisano, M., Vasi, F., Gerrish, P.J., and Schmidt, T.M. (1998) Evolution of competitive fitness in experimental populations of *E coli*: what makes one genotype a better competitor than another? *Antonie Van Leeuwenhoek Intl J General Mol Microbiol*, **73**, 35–47.
7. Lenski, R.E. and Travisano, M. (1994) Dynamics of adaptation and diversification: a 10,000-generation experiment with bacterial populations. *Proc Natl Acad Sci U S A*, **91**, 6808–6814.

8. Nielsen, J. (ed.) (2001) *Metabolic Engineering.* Springer, Berlin and Heidelberg.

9. Paquin, C.E. and Adams, J. (1983) Relative fitness can decrease in evolving asexual populations of *S. cerevisiae. Nature*, **306**, 368–371.

10. Zhang, Y.X., Perry, K., Vinci, V.A., Powell, K., Stemmer, W.P.C., and del Cardayre, S.B. (2002) Genome shuffling leads to rapid phenotypic improvement in bacteria. *Nature*, **415**, 644–646.

11. Gokhale, D.V., Puntambekar, U.S., and Deobagkar, D.N. (1993) Protoplast fusion— a tool for intergeneric gene-transfer in bacteria. *Biotechnol Adv*, **11**, 199–217.

12. Hopwood, D.A., Wright, H.M., and Bibb, M.J. (1977) Genetic-recombination through protoplast fusion in *Streptomyces. Nature*, **268**, 171–174.

13. Iwata, M., Mada, M., and Ishiwa, H. (1986) Protoplast fusion of *Lactobacillus fermentum. Appl Environ Microbiol*, **52**, 392–393.

14. Rassoulzadegan, M., Binetruy, B., and Cuzin, F. (1982) High-frequency of gene-transfer after fusion between bacteria and eukaryotic cells. *Nature*, **295**, 257–259.

15. Javadekar, V.S., Sivaraman, H., and Gokhale, D.V. (1995) Industrial yeast-strain improvement—construction of a highly flocculant yeast with a killer character by protoplast fusion. *J Ind Microbiol*, **15**, 94–102.

16. Petri, R. and Schmidt-Dannert, C. (2004) Dealing with complexity: evolutionary engineering and genome shuffling. *Curr Opin Biotechnol*, **15**, 298–304.

17. Alper, H., Moxley, J., Nevoigt, E., Fink, G.R., and Stephanopoulos, G. (2006) Engineering yeast transcription machinery for improved ethanol tolerance and production. *Science*, **314**, 1565–1568.

18. Alper, H. and Stephanopoulos, G. (2007) Global transcription machinery engineering: a new approach for improving cellular phenotype. *Metab Eng*, **9**, 258–267.

19. Liu, H.M., Yan, M., Lai, C.G., Xu, L., and Ouyang, P.K. (2010) gTME for improved xylose fermentation of *Saccharomyces cerevisiae. Appl Biochem Biotechnol*, **160**, 574–582.

20. McCullum, E.O., Williams, B.A.R., Zhang, J., and Chaput, J.C. (2010) Random mutagenesis by error-prone PCR. *Methods Mol Biol*, **634**, 103–109.

21. Runquist, D., Hahn-Hagerdal, B., and Bettiga, M. (2010) Increased ethanol productivity in xylose-utilizing *Saccharomyces cerevisiae* via a randomly mutagenized xylose reductase. *Appl Environ Microbiol*, **76**, 7796–7802.

22. Giaever, G., Chu, A.M., Ni, L., Connelly, C., Riles, L., Veronneau, S., Dow, S., Lucau-Danila, A., Anderson, K., Andre, B. et al. (2002) Functional profiling of the *Saccharomyces cerevisiae* genome. *Nature*, **418**, 387–391.

23. Winzeler, E.A., Shoemaker, D.D., Astromoff, A., Liang, H., Anderson, K., Andre, B., Bangham, R., Benito, R., Boeke, J.D., Bussey, H. et al. (1999) Functional characterization of the *S. cerevisiae* genome by gene deletion and parallel analysis. *Science*, **285**, 901–906.

24. Ross-Macdonald, P., Coelho, P.S.R., Roemer, T., Agarwal, S., Kumar, A., Jansen, R., Cheung, K.H., Sheehan, A., Symoniatis, D., Umansky, L. et al. (1999) Large-scale analysis of the yeast genome by transposon tagging and gene disruption. *Nature*, **402**, 413–418.

25. Goryshin, I.Y., Jendrisak, J., Hoffman, L.M., Meis, R., and Reznikoff, W.S. (2000) Insertional transposon mutagenesis by electroporation of released Tn5 transposition complexes. *Nat Biotechnol*, **18**, 97–100.

26. Jacobs, M.A., Alwood, A., Thaipisuttikul, I., Spencer, D., Haugen, E., Ernst, S., Will, O., Kaul, R., Raymond, C., Levy, R. et al. (2003) Comprehensive transposon mutant library of *Pseudomonas aeruginosa*. *Proc Natl Acad Sci U S A*, **100**, 14339–14344.

27. Sassetti, C.M., Boyd, D.H., and Rubin, E.J. (2001) Comprehensive identification of conditionally essential genes in mycobacteria. *Proc Natl Acad Sci U S A*, **98**, 12712–12717.

28. Kumar, A., Seringhaus, M., Biery, M.C., Sarnovsky, R.J., Umansky, L., Piccirillo, S., Heidtman, M., Cheung, K.H., Dobry, C.J., Gerstein, M.B. et al. (2004) Large-scale mutagenesis of the yeast genome using a Tn7-derived multipurpose transposon. *Genome Res*, **14**, 1975–1986.

29. Seifert, H.S., Chen, E.Y., So, M., and Heffron, F. (1986) Shuttle mutagenesis— a method of transposon mutagenesis for *Saccharomyces cerevisiae*. *Proc Natl Acad Sci U S A*, **83**, 735–739.

30. Badarinarayana, V., Estep, P.W., Shendure, J., Edwards, J., Tavazoie, S., Lam, F., and Church, G.M. (2001) Selection analyses of insertional mutants using subgenic-resolution arrays. *Nat Biotechnol*, **19**, 1060–1065.

31. Ni, H.Y., Laplaza, J.M., and Jeffries, T.W. (2007) Transposon mutagenesis to improve the growth of recombinant *Saccharomyces cerevisiae* on (D)-xylose. *Appl Environ Microbiol*, **73**, 2061–2066.

32. Wang, H.H., Isaacs, F.J., Carr, P.A., Sun, Z.Z., Xu, G., Forest, C.R., and Church, G.M. (2009) Programming cells by multiplex genome engineering and accelerated evolution. *Nature*, **460**, 894–898.

33. Warner, J.R., Reeder, P.J., Karimpour-Fard, A., Woodruff, L.B.A., and Gill, R.T. (2010) Rapid profiling of a microbial genome using mixtures of barcoded oligonucleotides. *Nat Biotechnol*, **28**, 856–U138.

34. Pierce, S.E., Davis, R.W., Nislow, C., and Giaever, G. (2007) Genome-wide analysis of barcoded *Saccharomyces cerevisiae* gene-deletion mutants in pooled cultures. *Nat Protocols*, **2**, 2958–2974.

35. Tyo, K.E.J., Ajikumar, P.K., and Stephanopoulos, G. (2009) Stabilized gene duplication enables long-term selection-free heterologous pathway expression. *Nat Biotechnol*, **27**, 760–765.

36. Lynch, M.D., Gill, R.T., and Stephanopoulos, G. (2004) Mapping phenotypic landscapes using DNA micro-arrays. *Metab Eng*, **6**, 177–185.

37. Lynch, M.D., Warnecke, T., and Gill, R.T. (2007) SCALEs: multiscale analysis of library enrichment. *Nat Methods*, **4**, 87–93.

38. Parekh, S., Vinci, V.A., and Strobel, R.J. (2000) Improvement of microbial strains and fermentation processes. *Appl Microbiol Biotechnol*, **54**, 287–301.

39. Minas, W., Sahar, E., and Gutnick, D. (1988) Flow cytometric screening 'and isolation of *Escherichia coli* clones which express surface-antigens of the oil-degrading microorganism *Acinetobacter calcoaceticus* Rag-1. *Arch Microbiol*, **150**, 432–437.

40. Wallberg, F., Sundstrom, H., Ledung, E., Hewitt, C.J., and Enfors, S.O. (2005) Monitoring and quantification of inclusion body formation in *Escherichia coli* by multi-parameter flow cytometry. *Biotechnol Lett*, **27**, 919–926.

41. Queener, S.W. and Lively, D.H. (eds.) (1986) *Screening and Selection for Strain Improvement*. Am Soc Microbiol, Washington, DC.

42. Berovic, M. and Legisa, M. (2007) Citric acid production. *Biotechnol Ann Rev*, **13**, 303–343.

43. Hermann, B.G., Blok, K., and Patel, M.K. (2007) Producing bio-based bulk chemicals using industrial biotechnology saves energy and combats climate change. *Environ Sci Technol*, **41**, 7915–7921.

44. Colin, T., Bories, A., Lavigne, C., and Moulin, G. (2001) Effects of acetate and butyrate during glycerol fermentation by *Clostridium butyricum*. *Curr Microbiol*, **43**, 238–243.

45. Kurian, J.V. (2005) A new polymer platform for the future—Sorona (R) from corn derived 1,3-propanediol. *J Polym Environ*, **13**, 159–167.

46. Otte, B., Grunwaldt, E., Mahmoud, O., and Jennewein, S. (2009) Genome shuffling in *Clostridium diolis* DSM 15410 for improved 1,3-propanediol production. *Appl Environ Microbiol*, **75**, 7610–7616.

47. Tatarko, M. and Romeo, T. (2001) Disruption of a global regulatory gene to enhance central carbon flux into phenylalanine biosynthesis in *Escherichia coli*. *Curr Microbiol*, **43**, 26–32.

48. Becker, J. and Boles, E. (2003) A modified *Saccharomyces cerevisiae* strain that consumes L-arabinose and produces ethanol. *Appl Environ Microbiol*, **69**, 4144–4150.

49. Wisselink, H.W., Toirkens, M.J., Wu, Q., Pronk, J.T., and van Maris, A.J.A. (2009) Novel evolutionary engineering approach for accelerated utilization of glucose, xylose, and arabinose mixtures by engineered *Saccharomyces cerevisiae* strains. *Appl Environ Microbiol*, **75**, 907–914.

50. Cakar, Z.P., Seker, U.O.S., Tamerler, C., Sonderegger, M., and Sauer, U. (2005) Evolutionary engineering of multiple-stress resistant *Saccharomyces cerevisiae*. *FEMS Yeast Res*, **5**, 569–578.

51. Kuyper, M., Toirkens, M.J., Diderich, J.A., Winkler, A.A., van Dijken, J.P., and Pronk, J.T. (2005) Evolutionary engineering of mixed-sugar utilization by a xylose-fermenting *Saccharomyces cerevisiae* strain. *FEMS Yeast Res*, **5**, 925–934.

52. Sonderegger, M. and Sauer, U. (2003) Evolutionary engineering of *Saccharomyces cerevisiae* for anaerobic growth on xylose. *Appl Environ Microbiol*, **69**, 1990–1998.

53. Guimaraes, P.M.R., Francois, J., Parrou, J.L., Teixeira, J.A., and Domingues, L. (2008) Adaptive evolution of a lactose-consuming *Saccharomyces cerevisiae* recombinant. *Appl Environ Microbiol*, **74**, 1748–1756.

54. Pang, Z., Liang, J., and Huang, R. (2010) Improvement of ethanol tolerance of xylose-fermenting yeast *Candida tropicalis* XY-19 by using genome shuffling. *Genomics Appl Biol*, **29**, 612–618.

55. Zheng, D.Q., Wu, X.C., Wang, P.M., Chi, X.Q., Tao, X.L., Li, P., Jiang, X.H., and Zhao, Y.H. (2011) Drug resistance marker-aided genome shuffling to improve

acetic acid tolerance in *Saccharomyces cerevisiae*. *J Ind Microbiol Biotechnol*, **38**, 415–422.

56. Bajwa, P.K., Pinel, D., Martin, V.J.J., Trevors, J.T., and Lee, H. (2010) Strain improvement of the pentose-fermenting yeast *Pichia stipitis* by genome shuffling. *J Microbiol Methods*, **81**, 179–186.

57. Dai, M.H. and Copley, S.D. (2004) Genome shuffling improves degradation of the anthropogenic pesticide pentachlorophenol by *Sphingobium chlorophenolicum* ATCC 39723. *Appl Environ Microbiol*, **70**, 2391–2397.

58. Zhao, H. and Chen, W. (2008) Chemical biotechnology: microbial solutions to global change. *Curr Opin Biotechnol*, **19**, 541–543.

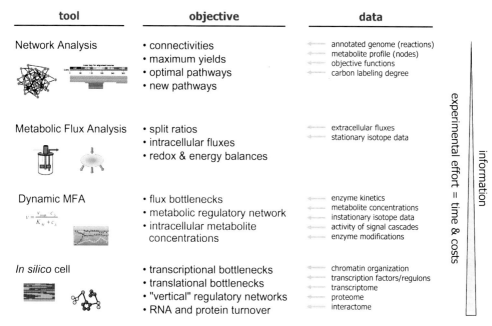

FIGURE 2.2. Metabolic engineering tools for the analysis of metabolic fluxes.

Engineering Complex Phenotypes in Industrial Strains, First Edition. Edited by Ranjan Patnaik.
© 2013 John Wiley & Sons, Inc. Published 2013 by John Wiley & Sons, Inc.

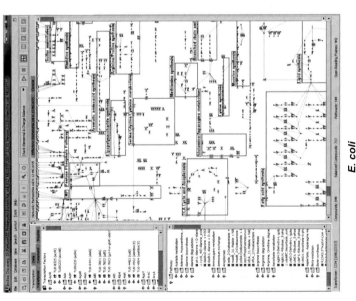

FIGURE 2.3. Genome-scale (A) and small-scale model (B) of *E. coli*. (See text for full caption.)

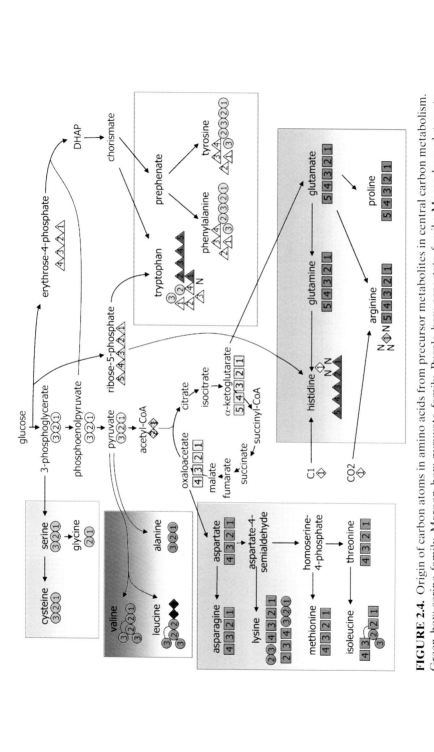

FIGURE 2.4. Origin of carbon atoms in amino acids from precursor metabolites in central carbon metabolism. Green box: serine family. Maroon box: pyruvate family. Purple box: aspartate family. Mocca box: glutamate family. Yellow box: shikimate family. DHAP, 3-deoxy-D-arabino-heptulosonate-7-phosphate; CoA, coenzyme A.

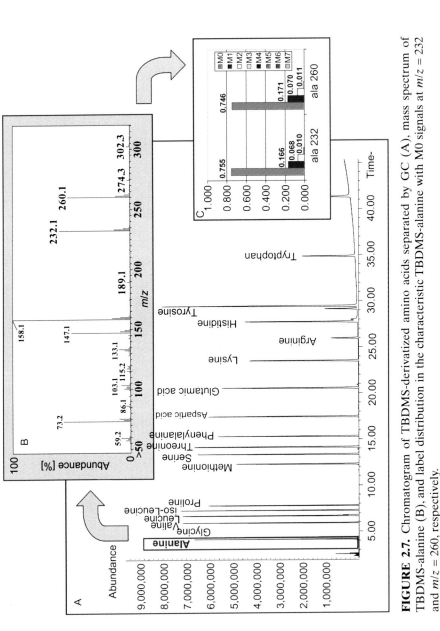

FIGURE 2.7. Chromatogram of TBDMS-derivatized amino acids separated by GC (A), mass spectrum of TBDMS-alanine (B), and label distribution in the characteristic TBDMS-alanine with M0 signals at $m/z = 232$ and $m/z = 260$, respectively.

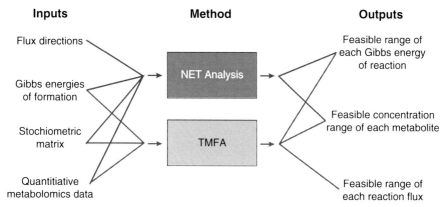

FIGURE 3.3. Comparison of network-embedded thermodynamic analysis (NET) and thermodynamics-based metabolic flux analysis (TMFA). Inputs and outputs to NET analysis and TMFA are color coded in red and blue, respectively.

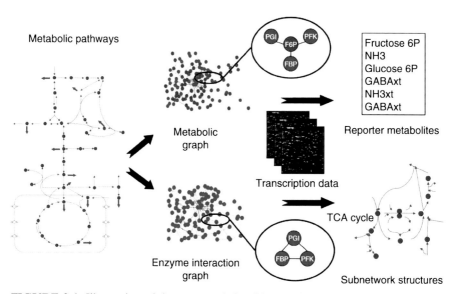

FIGURE 3.6. Illustration of the proposed algorithm for identifying reporter metabolites and subnetwork structures signifying transcriptionally regulated modules. A metabolic network is converted to metabolic and enzyme-interaction graph representations. Gene expression data from a particular experiment then are used to identify highly regulated metabolites (reporter metabolites) and significantly correlated subnetworks in the enzyme-interaction graph. TCA, tricarboxylic acid cycle; PGI, phosphoglucose isomerase; PFK, phosphofructokinase; FBP, fructose bisphosphatase. Reproduced from Reference (64), Copyright 2005 National Academy of Sciences, USA.

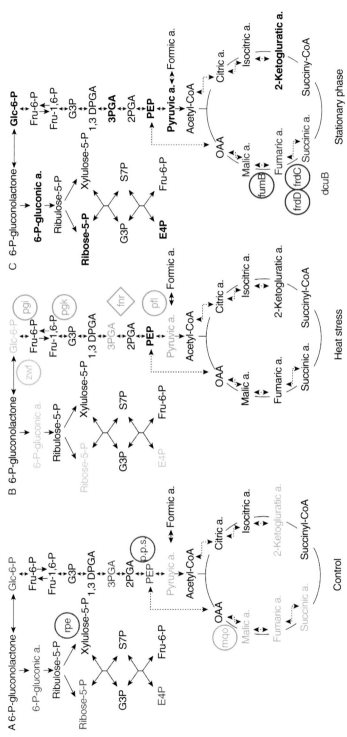

FIGURE 3.9. Canonical correlation analysis reveals condition-dependent association between response dynamics on the transcript and metabolite levels. Metabolites and genes showing close association were extracted and projected on a schematic representation of glycolysis, the pentose phosphate pathway, and the TCA cycle. Metabolic genes are circled and regulatory genes are displayed in a diamond box. Measured metabolites are indicated in bold. Transcripts and metabolites showing a close association are indicated by the same color. Reprinted by permission from Macmillan Publishers Ltd.: *Molecular Systems Biology* (87). Copyright 2010.

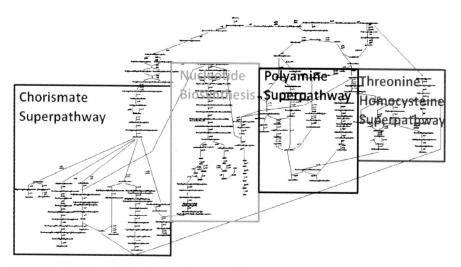

FIGURE 7.3. The 3-HP toleragenic complex (3-HP-TGC) as constructed from metabolic pathway fitness data. Subsections of the 3-HP-TGC are denoted for the chorismate, nucleotide biosynthesis pathway, polyamine, and threonine/homocysteine superpathways.

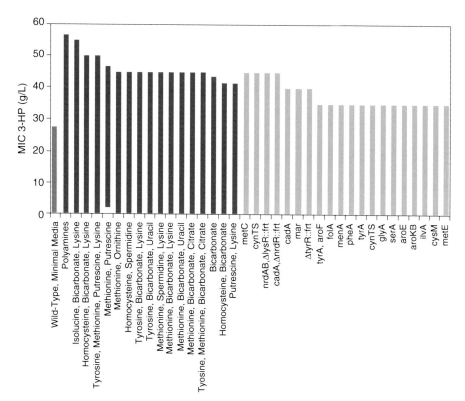

FIGURE 7.4. Confirmation of 3-HP tolerance corresponding to supplements (red) and genetic modifications (green). Tolerance was quantified as the minimum inhibitory concentration (MIC) of 3-HP in triplicate ($n = 3$) at pH = 7.0.

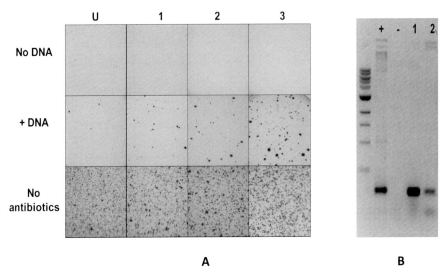

FIGURE 8.6. (A) Transformants of *Chlorella vulgaris*. U: untreated cells, 1–3: cell wall-degrading enzyme-treated cells. (B) PCR analysis of transformants. +: Vector DNA; –: untransformed gDNA; 1, 2: transformants gDNA.

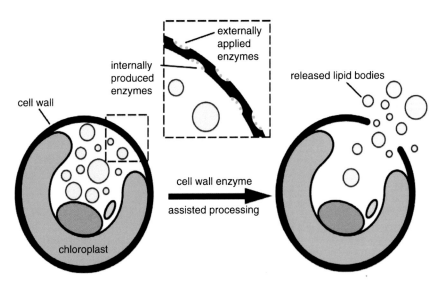

FIGURE 8.7. Release of internal algal oil bodies by internally or externally applied enzymes.

5

RAPID FERMENTATION PROCESS DEVELOPMENT AND OPTIMIZATION

Jun Sun and Lawrence Chew

5.0 INTRODUCTION

For the past decade advances in metabolic engineering and high-throughput technologies have made it possible to rapidly generate and screen large amounts of genetically engineered strains (see Chapters 1, 3, and 4) (1,2). The objective of these high-throughput technologies is to narrow down the list of strains to a very few (mostly just one) lead candidates that will be used in industrial-scale for production. Thus, challenges arise for bioprocess engineers to identify the best candidate strain that has the highest probability of demonstrating the desired performance on scale-up. These challenges include:

1. Effectively evaluating the performances of these large numbers of strains under conditions close to that observed in large-scale fermentors. The traditional strain evaluation strategy involves the processes of evaluating performance of strains from test tubes to shake flasks, then to benchtop fermentors, and eventually to pilot-scale fermentors. These processes are not only labor-intensive and time-consuming, but also costly if the starting pools of candidate strains are large.

2. Defining a set of criteria for the selection of the best candidate strain for scale-up. Product yield, rate, and titer are generally the most important parameters to justify strain performance. However, often these are not

Engineering Complex Phenotypes in Industrial Strains, First Edition. Edited by Ranjan Patnaik.
© 2013 John Wiley & Sons, Inc. Published 2013 by John Wiley & Sons, Inc.

the only factors to be considered. Sometimes process operability and downstream process constraints will impose a big impact on total production cost, necessitating further analysis of strain performance in the light of additional metrics. For example, a strain with the highest yield, rate, and titer may not be a winner if it causes unreliable fermentation process and unrealistically high downstream purification cost. Some of these additional performance criteria can be addressed either by engineering approaches or more cost-effectively by engineering these traits into production strains.

3. Rapidly developing an optimized robust process for scale-up after strains are identified from high-throughput screening or selection programs. A variety of scale-down tests may need to be performed at this stage to address potential problems that might surface at large scale, such as mass and/or heat transfer limitations resulting from high oxygen utilization rates of the culture.

This chapter will illustrate a roadmap on how to use new fermentation tools to address the above three challenges in the most cost-effective and practical manner such that the throughput of a fermentation process development program is compatible with the throughput of typical strain development programs.

5.1 OVERVIEW OF CLASSICAL FERMENTATION PROCESS DEVELOPMENT METHODOLOGY

The traditional fermentation process development workflow can be illustrated in Figure 5.1A. First, a pool of strains identified from strain screening or metabolic engineering efforts are tested in either test tubes or shake flasks. By comparing the final product titer and yield indicative of each strain, the size of the initial strains pool is reduced for a more detailed characterization. This process can be performed iteratively in several cycles with varied culture conditions and media compositions for each cycle. Typically, the number of the selected strains that are moved forward for further characterization in fermentors is limited by the capacity of the fermentation laboratory. It is difficult to quantitatively monitor and control all important scale-up parameters such as pH, dissolved oxygen (DO) level, and substrate concentration profiles in test tubes or shake flasks, as typical results from flask characterization only provide an end-point data in batch fermentation mode. However, these results from flask characterization are sufficient to compare the performance of the strains if the final production process is a simple anaerobic batch process, such as ethanol fermentation. For fed-batch production process and other aerobic fermentations where oxygen transfer rates are important for scale-up, this type of screening method may miss some potential good candidates due to limitations arising from the batch mode of operation.

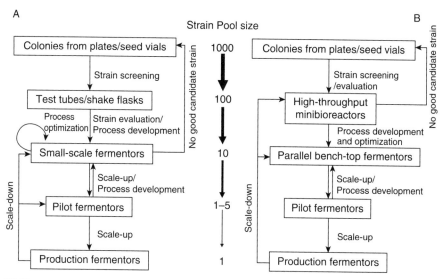

FIGURE 5.1. Roadmap for traditional fermentation process development (A) and for rapid fermentation process development using high-throughput fermentation systems (B).

The next step is to evaluate each selected strain from the preliminary screening stage under well-controlled conditions in small-scale fermentors (1–20 L). This is probably the most laborious and expensive step during process development. Some statistical methods and mathematical modeling tools can be used to design fermentation experiments to evaluate the strains under a broad range of conditions that are unable to be controlled in shake flasks. After extensive strain evaluation, a few strains will be selected based on product yield, rate, and titer for further process development and optimization. Many times it will end up with no good candidate strain to move forward, and the process returns to the first step for screening more strains as illustrated in Figure 5.1A.

When a process is developed at small scale, it will take a couple of tests to validate the performance at different pilot scales. Further optimization of the process is required to scale-up to production scale. At this stage, many scale-down experiments should be performed to address the potential problems that may occur at production scale. These tests include strain stability and process robustness subjected to the constraints of the large scale. However, it is not unusual that some specific problems will be found only after the process is scaled up. In such cases more scale-down experiments will be designed to operate at small-scale fermentors to identify the causes of the problems and develop possible solutions.

The above-described traditional approach for industrial strain development and scale-up puts significant pressure on fermentation equipment resources at the 1–20 L scale, and often this step becomes the bottleneck in a strain development program. In order to improve the throughput of strains that are characterized in representative fermentation conditions, in recent years new technologies on high-throughput fermentation have been developed to keep pace with the throughput of the strain engineering programs.

5.1.1 Noninvasive Sensor Technologies

The traditional process development strategy as described in Section 5.1 can last for years and cost tremendous resources to the strain engineering program. The bottleneck of the process is the speed of strain evaluation and process development. To alleviate the bottleneck, there are two approaches: (i) change the screening methods to evaluate the strains more effectively such that false positives are minimized, and (ii) increase the throughput of fermentation at the small scale without significantly compromising the metrics relevant to scale-up.

Fortunately, recent development in new fermentation technologies makes these approaches possible. Thanks to the noninvasive sensing technologies, it becomes practical to attach optical pH and DO sensors to the flasks that enable monitoring and controlling pH/DO in fermentation cultures grown in shake flasks. Furthermore, when these optical sensor technologies are integrated with microwell reactors, it creates minibioreactors that have most of the functionalities and similar performances to that for small benchtop fermentors, plus the high-throughout capability. This leads to a new roadmap for rapid strain evaluation and process development as proposed in Figure 5.1B.

In this case the minibioreactors will combine the functions of both shake flasks and small-scale fermentors (as shown in Figure 5.1A) to perform the strain screening and evaluations. An early decision can be made if there is no good strain for further process development. The high-throughput fermentors can expedite the process development by enabling optimization across a wider range of operating parameters. In addition, some scale-down experiments reflective of the scaled-up process can even be done in minibioreactors without the need for piloting, which results in significant cost savings. By using these approaches, the bottleneck in traditional fermentation process development can be dramatically alleviated, although not completely eliminated.

In this chapter, the basic tools for fermentation process development will be introduced and illustrated by some case studies. The advances in minibioreactor technologies will be reviewed, and the different options of commercially available minibioreactors and parallel benchtop high-throughput fermentation systems will be introduced and compared. Some case studies will be presented to demonstrate how to use the available high-throughput fermentation systems to expedite strain evaluation and process development.

5.2   FERMENTATION PROCESS DEVELOPMENT AND OPTIMIZATION

Fermentation process development and optimization includes optimization of growth medium and growth conditions to achieve maximal economic advantage at production scale. It also involves optimization of fermentation performance using a scale-down approach to simulate the conditions typically encountered at scale-up. This section will discuss the development and optimization of microbial fermentation processes using currently available technologies.

5.2.1   Medium Design and Optimization

When evaluating strains and developing a fermentation process, the fermentation medium is one of the most critical factors that influences strain performance. The medium design and strain screening process often form an iterative cycle to test the strain pools in different media in order to increase the probability of finding an optimal combination. It is impossible for one medium to fit or be conducive for all strains.

There are typically two strategies for medium design and optimization: (i) "open strategy," which involves selecting the best combination of all possible components, and (ii) "closed strategy," which is done by identifying the best combination of given components. The different approaches on how to design or optimize media using the above two strategies are summarized in References (3) and (4).

5.2.1.1   Phenotype Microarray for Rapid Media Development   Component swapping (swap one component for another at the same level) (5,6) and pulsed injection in continuous fermentation techniques (addition of growth limiting nutrient results in growth simulation) (7,8) are two experimental techniques for medium design with open strategy. Combined with mathematical and statistical tools, these two methods have been successfully applied to identify the essential medium components for microbial fermentation. However, a single-medium optimization study using these low-throughput "trial and error" methods will take a long time, with high variation in results. They are not suitable for medium design when a significant number of strains generated by genetic engineering are being evaluated because the genetically engineered strains will have different requirements for optimal fermentation performance. To determine the best medium composition for the specific strain or most promising media for strain screening, a systematic physiology-driven approach is preferable.

The recent advances in metabolomics have made it possible to design a medium using an unbiased selection and ranking in high-throughput formats (9). For example, the phenotype microarray (PM) from Biolog

(Hayward, CA) is a high-throughput method to globally characterize cell phenotype and reveal microbial cell physiology via a systematic approach (10,11). The PM set for bacteria consists of 1920 phenotype assays in 96-well microplates to test cell metabolism and chemical sensitivity (12). The microtiter plate wells contain chemicals dried on the bottom to create unique culture conditions after rehydration. Assays are initiated by inoculating all wells with cell suspensions. After incubation, some of the wells turn various shades of purple due to reduction of a tetrazolium dye as the cells respire. Instead of measuring cell growth, cell respiration activities are measured colorimetrically to give an accurate reflection of the physiological state of cells with improved sensitivity of assays. This is particularly important for some assays that do not depend on growth. Assays of carbon (C), nitrogen (N), phosphorous (P), and sulfur (S) metabolism provide quantitative information about the activity of various metabolic pathways that are present and active in cells. Assays of ion, pH, and chemical sensitivities provide information on stress and repair pathways that are present and active in cells.

By using the PM system, it is feasible to compare and fingerprint differences between the genetically engineered strains generated in a strain engineering program. Furthermore, combining with statistical tools such as multivariate data analysis, it is possible to scan 2000 culture conditions simultaneously to identify the essential growth medium components. Knowledge and insights obtained from phenotypic fingerprinting of strains not only influence cost-effective medium design but also enable modulating other traits through either addition or exclusion of certain micronutrients.

5.2.1.2 *Media Optimization Using Design of Experiment (DOE)* The traditional method for medium development is a trial–error process in shake flasks or benchtop fermentors. Typical approaches include changing one variable at a time to determine the impact of one component on the strain performance. But this approach cannot identify the interactive effect of multiple components on process performance.

The DOE approach to optimize medium formulation has been used widely by microbiologists and fermentation engineers. The most commonly used DOE methods include:

1. Factorial design methods allow for the simultaneous study of the effects that several medium components may have on performance. Varying the levels of many components simultaneously rather than one at a time allows for the study of interactions between the components. Using the factorial design, a few vital components can be quickly identified for further optimization. The fractional two-factorial Plackett–Burman design, which allows for investigation of up to N-1 variables in N experiments, is an efficient design used frequently as a starting point when many variables need to be screened to identify the vital components to be optimized further.

2. Response surface methods are used to examine the relationship between one or more fermentation performance variables (e.g., product yield, rate, and titer) and a set of "vital few" medium components identified by other methods (e.g., factorial design methods). The medium components can be optimized to achieve best fermentation performance. Designs of this type are usually chosen when the curvature in the response surface may exist.

Canonical analysis is a method used to further characterize the surface response function to identify the minimal or maximal points on the response surface. In most cases, it is difficult to understand the shape of a fitted response surface by mere inspection of the algebraic expression of the Taylor polynomial. When there are many independent variables in the model, it is also difficult to evaluate the shape of the surface by looking at isocontour projections of the variables two by two. A canonical analysis facilitates the interpretation of the results obtained by surface response methods (13). A detailed example using above three statistic tools to optimize the medium components for nattokinase production by *Bacillus natto* NLSSEm can be found in Reference (14). Many commercial software, such as Statistica (Statsoft Inc, Tulsa, OK) and Minitab (Minitab Inc., State College, PA) can be used to facilitate DOE and data analysis.

When selecting a medium for fermentation, it has to be kept in mind that most often the final product will need to be separated from the medium components at the end of the process. Therefore, the medium used at production scale should be as lean as possible not only to meet the requirement of cell growth and product formation but also to minimize the medium cost and downstream processing burden (see case study in Chapter 6).

5.2.2 Optimization of Growth Conditions

The use of DOE methods as described in Section 5.2.1.2 has also been reported extensively to optimize the growth conditions (15). Some conditions such as pH, temperature, and induction/fermentation batch cycle time can be controlled in the shake flasks. Thus, it is possible to optimize these growth conditions in shake flasks or microtiter plates. However, some growth parameters such as DO cannot be easily measured and controlled in the shake flasks. Thus, the small-scale fermentor is a preferable tool.

When using fermentors to optimize growth conditions, it is laborious to run 10–20 batches of fermentation based on DOE. One traditional approach is to use continuous chemostat culture to test different growth conditions at each steady state. However, depending on the growth rate, it usually takes a long time to reach a steady state in continuous culture. To test growth conditions at different levels, the continuous fermentation may last for a long period, imposing the risk of contamination or genetic instability. With the combination of mathematical modeling, it becomes practical to identify optimal growth conditions by varying the growth parameters before a steady state is reached (16,17).

5.3   RAPID PROCESS DEVELOPMENT AND OPTIMIZATION USING CONVENTIONAL FERMENTATION SYSTEM

The fed-batch fermentation is the most widely used fermentation mode to achieve high product titer and yield. When developing fed-batch fermentation, the most important variable that needs to be defined is the substrate feed profile. Various methods have been described to determine and optimize the feed profile for specific strain and process as reviewed by Lee in Reference (18). However, when evaluating multiple strains to identify the good candidates, it is impractical to determine the optimized feed profile for each strain. A common strategy is to use a predefined feed profile based on a designated specific growth rate for all strains to be evaluated. This may exclude some good candidates because the predefined feed profile may not be optimal for these strains. Thus, it is preferable to define and customize feed profile online during the fermentation process for each strain to be evaluated. The following two methods are practical and robust to define the feed profiles for carbon source-limited fermentation or carbon source-excess fermentation.

5.3.1   Dynamic DO Control to Determine Optimal Feed Rate for Carbon Source-Limited Fermentation

Many fed-batch fermentation processes need to keep carbon source at a limited level to minimize by-product formation. For example, when producing brewery yeast using fed-batch fermentation, excess sugar in the culture leads to the production of ethanol. Another example is *Escherichia coli* fermentation for protein expression, where excess glucose in the media can divert the carbon flux into acetate formation, resulting in inhibition of cell growth and protein expression.

When a fermentation system is equipped with advanced instrumentations, such as online measurement of substrates using bioanalyzer or off-gas measurement by gas analyzer, and so on, it is not difficult to design a feedback program to control the feed rate to evaluate strains with different growth phenotypes. However, most small-scale fermentors are equipped only with basic controllers for pH, DO, temperature, and so on, but not any advanced instrumentation. Thus, a practical approach is needed to define feed profile using a standard fermentation system without additional advanced instrumentation.

For strain evaluation purposes, the feeding strategy should be (i) universal (independent of strains, products, media, etc.), (ii) robust, (iii) requiring mínimum number of online measurements as possible, (iv) amenable to scale-up, and (v) ease of implementation, preferably with full automation. Based on these criteria, DO-stat and pH-stat are the two best choices as DO and pH probes are relatively reliable and cheap. Most of the modern fermentation systems have the DO-stat/pH-stat substrate feeding control program built in the control unit. But because the pH-stat has several disadvantages such as slow response to feed addition and media dependence, the DO-stat is a suitable strategy for substrate limited fed-batch fermentation for strain evaluation.

The commonly used DO-stat control strategy for fed-batch fermentations involves adding substrate when the DO level rises above the setpoint due to substrate depletion. When this strategy is used for certain fermentations such as recombinant protein expression or when the product is toxic to the cells, it can cause substrate overfeeding as shown in Figure 5.2A (19). In this example, a built-in DO-stat control program in a Sartorius Biostat C control unit (Sartorius, Göttingen, Germany) was used to control feed rate of the glucose feed solution by keeping the DO value at 30%. However, after induction at 8.5 hours, the DO stayed higher than its setpoint due to slowing down of cell metabolism, subsequently resulting in overfeeding of glucose. The glucose feed has to be shut off manually to prevent further adverse impact of high glucose concentration on the protein expression.

When using DO-stat, one approach to resolve the overfeeding problem is to feed the substrate dynamically by the pattern of DO change. As described in Reference (19), the increase in DO can be caused by (i) substrate depletion; (ii) slowing down of cellular metabolism or product toxicity; (iii) noise of DO probe. If standard DO-stat control strategy is used, substrate over-feeding will occur for scenarios of (ii) and (iii). Fortunately it is possible to design a software controller to distinguish the patterns of DO change and to feed substrate only in response to substrate depletion.

A control strategy of differentiating the pattern of DO change by changing the DO setpoint dynamically to follow the DO profile with 1–2 minutes of delay has been described in detail in Reference (19). In this dynamic DO-stat feeding control strategy, the feed will be triggered if DO is higher than DO setpoint (DO_{set}) plus a high threshold value (DO_{ht}) which is a value higher than the probe noise. When substrate is exhausted, the DO will increase drastically and the feed will start because $DO > (DO_{set} + DO_{ht})$. In the case of the cell metabolism slowing down or product toxicity, the DO will increase slowly and the DO setpoint will dynamically follow the DO profile with 100s of delay to avoid substrate overfeeding. For the DO probe noise issue, the parameter of DO_{ht} which is higher than the probe signal noise will prevent substrate overfeeding.

This dynamic DO-stat can be easily programmed and has been implemented using different commercially available fermentation SCADA software (19). An example of fed-batch fermentation using this dynamic DO-stat

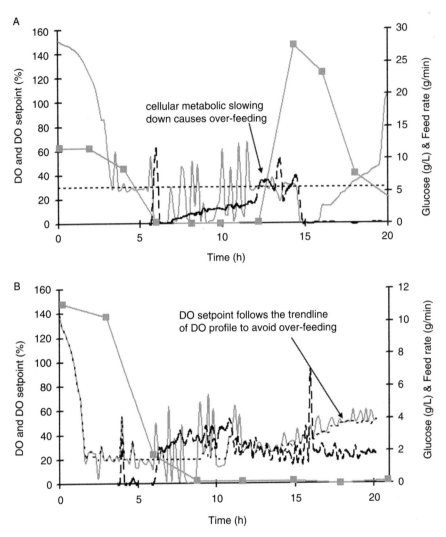

FIGURE 5.2. Examples of fed-batch fermentation using conventional DO-stat (A) or dynamic DO-stat (B) to control the substrate feed rate. In both fermentations, a recombinant protein was expressed in *E. coli* cell. The feed solution is 50% (w/w) glucose. The protein expression was induced at 8.5 hours. The DO setpoint is at 30% for (A). The minimal DO setpoint is at 20% for (B). Grey solid line: DO profile (%); dashed line: feed rate (g/min); grey dotted line: DO setpoint (%); square: glucose concentration (g/L). Part (A) is adapted with kind permission from Springer Science + Business Media: In Cheng, Q. (ed.), *Microbial Metabolic Engineering, Methods in Molecular Biology*. Chapter 15, Rapid strain evaluation using dynamic DO-stat fed-batch fermentation under scale-down conditions. SpringerLink, New York, vol. **834**, 2012, pp. 233–244, Jun Sun.

control strategy is shown in Figure 5.2B. After induction at 8.5 hours, the cell metabolism slowed down and caused DO to increase above 20%, but the control program was able to distinguish this DO increase pattern and controlled glucose feeding accordingly to keep the glucose at limited level. Another advantage of using this dynamic DO-stat is that the by-product can be reduced for *E. coli* fermentation as reported in Reference (20) because of the oscillation of DO.

The above dynamic DO-stat control strategy is suitable for initial strain screening and process development without prior knowledge about the substrate utilization rate or growth rate of strains. It can provide a baseline feed profile after strain evaluation for further process optimization.

5.3.2 Feed Forward Control for Carbon Source Excess Fermentation

The feed control for substrate excess fermentation is not as challenging as for substrate-limited fed-batch fermentation because the substrate concentration will be controlled within a certain range. However, when the substrate level is out of the specified range, the fermentation performance will be impacted adversely due to by-product formation or substrate inhibition. Among the various feed strategies as described in Reference (18), the direct feed back control based on substrate concentration is suitable for substrate excess fermentation and is easy to implement. Since the substrate level only needs to be controlled within certain range, most of the time no online measurement of substrate is needed. A rapid offline measurement of substrate level with 2–3 hours of sampling interval is generally enough to meet the requirement for feed rate adjustment. One option is shown in Table 5.1. In this case, the online or offline measurement for broth weight, cumulative feed weight, and substrate concentration is used to calculate the substrate consumption rate. Then the substrate consumption rate for the next sampling interval is calculated by linear extrapolation of the substrate consumption rate of the past two sampling intervals. Thus, the feed rate can be calculated based on the predicted substrate consumption rate plus a term to calibrate the difference between substrate concentration and the setpoint for substrate level. This method can be implemented in spreadsheet software or in the modern fermentation SCADA software for automatic control of the feed rate. One example of using this feed forward control strategy is shown in Figure 5.3. A sample was taken every 2–3 hours and measured for sugar concentration. The offline measurement was entered into an excel file to predict the feed rate for next sampling interval. The robustness of this method can be demonstrated by the sugar level, which was stably controlled within the range of 10–15 g/L with the setpoint of 15 g/L. This method is suitable for strain evaluation when no prior knowledge of strain growth property is available and excess substrate during fermentation is desirable. It is also very useful for process development and scale-up since the method is independent of scale.

TABLE 5.1. A Simple Feed Forward Control Algorithm to Predict the Feed Rate Based on Offline Measurement of Broth Weight (W), Substrate Concentration (S), and Cumulative Feed Weight (CF). The Substrate Concentration in Feed Solution Is S_{in} (g/g). Substrate Level Setpoint Is S_{set} (g/g)

Time (hr)	Broth Weight (g)	Cumulative Feed Weight (g)	Substrate Concentration (g/g)	Substrate Consumption Rate σ (g/h)	Predicted Feed Rate for the Next Sampling Interval (g/h)
t_0	W_0	CF_0	S_0	$\sigma_0\ (=0)$	0
t_1	W_1	CF_1	S_1	$\sigma_1 = ((CF_1 - CF_0)S_{in} + W_1 S_1 - W_0 S_0)/(t_1 - t_0)$	$\sigma_1(\sigma_1(t_2 + t_1 - 2t_0) - \sigma_0(t_2 - t_1))/((t_1 - t_0)(\sigma_1 + \sigma_0)S_{in}) - W_1(S_1 - S_{set})/S_{in}/(t_2 - t_1)$
t_2	W_2	CF_2	S_2	$\sigma_2 = ((CF_2 - CF_1)S_{in} + W_2 S_2 - W_1 S_1)/(t_2 - t_1)$	$\sigma_2(\sigma_2(t_3 + t_2 - 2t_1) - \sigma_1(t_3 - t_2))/((t_2 - t_1)(\sigma_2 + \sigma_1)S_{in}) - W_2(S_2 - S_{set})/S_{in}/(t_3 - t_2)$
t_3	W_3	CF_3	S_3	$\sigma_3 = ((CF_3 - CF_2)S_{in} + W_3 S_3 - W_2 S_2)/(t_3 - t_2)$	$\sigma_3(\sigma_3(t_4 + t_3 - 2t_2) - \sigma_2(t_4 - t_3))/((t_3 - t_2)(\sigma_3 + \sigma_2)S_{in}) - W_3(S_3 - S_{set})/S_{in}/(t_4 - t_3)$
t_4	...	...	...	...	...

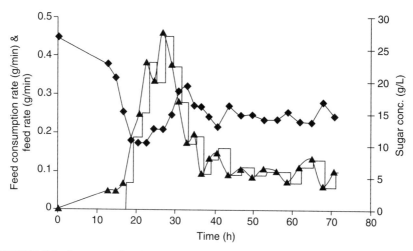

FIGURE 5.3. An example of a bacterial fed-batch fermentation using sugar as substrate. The feed forward control program as shown in Table 5.1 was used to generate the feed profile. Diamond: sugar concentration; triangle: calculated feed consumption rate as described in the 5th column in Table 5.1; solid line: actual feed rate predicted by the 6th column in Table 5.1.

5.4 STRAIN EVALUATION AND PROCESS OPTIMIZATION UNDER SCALE-DOWN CONDITIONS

While fermentation in small-scale fermentors can be a homogeneous and well-controlled process, most of the time it does not match the conditions in large-scale fermentors, where the gradient of pH, temperature, DO, and substrates can be significant. Furthermore, there are mass and heat transfer capacity limitations at larger scale that further constraints fermentation optimization. Cells in a large-scale reactor move through different zones and experience steady changes in their environment (21). Consequently, they experience a different history with respect to their physiology, especially stress responses compared with cells that have been grown in a homogeneous well-mixed culture (22). This often significantly affects the fermentation performance in large vessels. Therefore, when using small-scale fermentors for strain evaluation, the performance of the leading candidate may not translate to large-scale vessels. It is often necessary to test certain scale-down conditions during evaluation of the candidate strains so that the performance of the best strain translates to a large-scale vessel. The review paper by Neubauer and Junne (21) has given the latest update on how to set up the systems to simulate the conditions in large-scale reactors using conventional benchtop fermentors, phototrophic fermentors, and micro high-throughput bioreactors, as well as study the metabolic changes with rapid sampling techniques and computational tools.

5.4.1   Identify Scale-Down Parameters

To simulate the conditions of a large-scale reactor in small-scale vessels, the right parameters should be identified so that the scale-down fermentation system can be set up accordingly.

The leading cause of deviation of fermentation performance in large-scale vessels from small vessels is the heterogeneity due to the limitation of power input for mixing in large-scale vessels. This leads to gradient formation for pH, substrate, temperature, and a hot spot in the large vessel, into which the substrate and pH adjusting agents are fed. It also causes the limitation of mass transfer, especially oxygen transfer rate (OTR) for aerobic fermentation. In small-scale fermentors, the OTR can sometimes reach as high as 300 mM/h using air as the oxygen supply, in contrast to about 100–150 mM/h in large-scale vessels. In addition, because of the hydrostatic pressure and good solubility of CO_2 at high CO_2 partial pressure, the dissolved CO_2 level in large-scale vessels is generally higher than that in small-scale vessels. Certain strains cannot tolerate high dissolved CO_2, which indirectly affects performance. If these scale-down parameters can be tested at the strain evaluation and process development stage, some engineering solutions can be implemented to resolve these issues at the scale-up stage. In this section, practical approaches for testing these scale-down parameters in lab-scale fermentors are discussed.

5.4.2   Scale-Down of Mixing Related Parameters

One powerful tool for studying the effect of mixing-related scale-down parameter is using the multiple-component fermentation system. The setups were summarized by Neubauer and Junne (21). These setups have been successfully applied to study the impact of gradients in substrate feeding zone (23,24) and pH agent feeding zone (25,26) on fermentation performance. The dynamic DO-stat control (19) as discussed in Section 5.3.1 is actually the pulse feeding scale-down strategy.

Except for the setup of direct rapid sampling and pulse feeding, which can be implemented in commercially available bioreactors, other configurations, such as stop-flow sampling and two compartment reactors, require setting up of multiple fermentors (21). This limits the application of these scale-down strategies during the early stages of strain evaluation due to the cost of labor and equipment. A high-throughput format of the above-described configurations using microliter bioreactors is more suitable to test the scale-down parameters as described in Section 5.7.

5.4.3   Oxygen Uptake Rate (OUR) Clipping

In large-scale vessels, the OTR is limited to 100–150 mM/h due to limitations on mixing power supply and volumetric air flow rate (vvm). In benchtop scale fermentors, the OTR can easily reach more than 200 mM/h. Thus, it is necessary to test the performance of engineered strains under the OTR limiting conditions early at the strain evaluation stage. Since OUR equals OTR at pseudo-steady state, this scale-down strategy is called OUR clipping.

The most common method of testing OUR clipping is to limit the maximal feed rate to keep the OUR below the setpoint. If a gas analyzer is available for online off-gas monitoring, it is easy to set the maximal feed rate based on online calculation of OUR. However, if the fermentor is not equipped with an off-gas analyzer, the approach can be as follows: (i) measure oxygen transfer coefficient (k_{La}) and correlate k_{La} with agitation speed and air flow rate (27); (ii) set the maximal agitation speed and/or air flow rate based on their correlations with k_{La} so that the maximal OTR can match that in the large-scale vessel; and (iii) design the feed profiles to limit the maximal feed rate when agitation speed and air flow rate reach their maximal setpoints.

An example of comparison of fermentation performance with and without OUR clipping is described in Reference (19). In the OUR clipping experiment, the maximal agitation speed and air flow were set to the current process values when OUR reached the setpoint. It was observed that cells grew slightly slower at OUR clipping condition (OUR was set at 165 mM/h) compared to regular fed-batch fermentation where OUR reached as high as 300 mM/h. The feed profiles for both runs were generated automatically using dynamic DO-stat as discussed in Section 5.3.1.

5.4.4 Dissolved CO_2

The dissolved CO_2 level in large-scale vessels is 2–3 times higher than that in the benchtop fermentor because of the higher hydrostatic pressure and low volumetric air flow rate (vvm) in large-scale vessels. It has been reported that high dissolved CO_2 (or high partial CO_2 pressure) can impose a negative impact on fermentation performance (28–31). Thus, it is important to check the fermentation performance on high dissolved CO_2 level at small scale before scale-up.

The most common approach to evaluate the effect of CO_2 on fermentation performance is using mixed gas with high CO_2 concentration as influent gas (29,30). This can mimic the conditions observed in large-scale fermentors with elevated CO_2 level. Another approach is autogenous CO_2 methods in which the airflow is adjusted automatically to keep off-gas CO_2 at the desired level (31). If the small fermentor is equipped with a pressure regulator as that in many sterilize-in-place (SIP) fermentation systems, the dissolved CO_2 effect can be evaluated by adjusting vessel back pressure to keep partial CO_2 pressure at a preset level equivalent to that in large-scale vessels.

5.5 CONTROL AND SENSOR TECHNOLOGIES FOR MINIBIOREACTOR

As discussed in Sections 5.2 to 5.4, the regular benchtop fermentors (0.5–10 L) are powerful tools to evaluate strains and develop scalable processes. However, as the pool size of the strains to be evaluated gets large, it is unrealistic to use benchtop fermentors to accomplish the strain screening and evaluation task effectively due to the labor, time and cost to operate benchtop fermentors. While many labs are still using the shake flasks or microtiter plates to screen the strains, the minibioreactors (50 μL–200 mL) equipped with temperature, pH, DO, and aeration sensors/controllers can offer high throughput with high reproducible and comparable results as that obtained from regular benchtop fermentors. The high data density from minibioreactors enables thorough data analysis to select and rank strains effectively using multiple metrics.

The minibioreactor can be classified in two categories based on the reactor volume: milliliter bioreactor (ranging from 1 to 200 mL) and microliter bioreactor (ranging from 5 to 700 μL). As the volume decreases, losses due to water evaporation become a challenging issue. From the viewpoint of practical application, the milliliter bioreactor can offer more flexibility in terms of the amount of sampling allowed and commercial availability.

The key factor in successful application of minibioreactor in strain screening and process development programs is the integration of sensor technologies with the hardware of the equipment. Because of the volume limitation of minibioreactors, online measurement of many fermentation parameters implemented in minibioreactors can prevent significant loss of culture broth due to sampling. In this section, sensing and control technologies used in minibioreactors are reviewed. Most of the sensors for temperature, pH, DO, and biomass have been integrated together into different minibioreactors for practical applications (32,33).

5.5.1   Temperature Sensing and Control

In minibioreactors, temperature is typically measured by thermistors or resistance temperature detectors (RTDs). These sensors are commercially available and can be fabricated in small sizes to be embedded into minibioreactors to measure temperature precisely and reliably.

While it is relatively easy to measure temperature, it is a challenging task to control the temperature because of the high surface area to working volume ratio (S/V). The smaller the working volume, the higher is the S/V ratio, thus the faster is the heat transfer or heat loss from the reactors. Therefore, it is necessary to have a well-functioning temperature control loop in minibioreactors to keep the temperature precisely at the desired setpoint.

Based on the type of minibioreactors, various methods have been applied to control temperature. For most of commercial minibioreactors, incubator or heated chamber are used to control temperature. However, for microliter reactors different approaches have been used to regulate temperature such as heating foil (34) or resistive heating (35) mounted on the base of microreactor. This can allow tight control of temperature. It is preferred to have an integrated heater in microliter bioreactors for temperature control because of the cost and flexibility of allowing parallel operation of individual bioreactors at different temperatures.

5.5.2 Mixing

A bioreactor has good mixing properties if it meets the following criteria: (i) it maintains the homogeneity of fermentation broth, which means short mixing time, and (ii) it provides enough power for mass/heat transfer but does not generate high shear force to disturb microbial physiology. As the scale of the reactor decreases, it is more difficult to provide adequate mixing power because of a small Reynolds number, which is an indication of turbulence. Based on the type of minibioreactors as summarized in Reference (36), the mixing power is provided by (i) shaking in the case of microtiter plate and shake flask-type minibioreactors; (ii) stirring in the case of stirred minibioreactors; and (iii) microfluidics in the case of microliter bioreactors embedded on chips (34).

Among the three mixing methods, the stirring method provides enough mixing power and support OTR as close as attainable in benchtop fermentors. The power consumption and k_{La} for different minibioreactors have been discussed in References (27) and (37).

5.5.3 DO

Oxygen is a critical variable to be monitored and controlled during aerobic fermentation process. For past decades, the DO level is measured mostly by Clark-type oxygen electrode which has serious drawbacks and limitations on the application for minibioreactor, such as its bulk size, signal to noise ratio and signal drift. For minibioreactor, the optical DO sensor (optode) based on the quenching of fluorescence by oxygen (38,39) is a better option and has been widely used in commercial minibioreactors and other microreactor prototypes. The DO optode can be easily embedded into the minireactor as a disposable DO sensor. Another option of DO sensor used in minibioreactors is electrochemical sensor such as ultra-microelectrode array (UMEA), an amperometric sensor measuring DO based on the electrochemical reduction of oxygen. This kind of sensor can be fabricated small enough to fit into individual well of a 96 microtiter plate (33).

The control of DO in minireactors is a critical task because of the difficulty of providing enough mixing power to promote mass transfer as described in Section 5.5.2. For shake flask type and microwell type minibioreactor, the shaking speed is the only way to control the OTR. However, for certain milliliter bioreactor, aeration can be used to control the DO as in μ-24 microreactor (Pall Corporation, Covina, CA). For stirred minibioreactors, varying the stirring speed can be much more effective than aeration to control DO. Stirring in combination with aeration systems such as gas-inducing impellers (40) can generate k_{La} as high as 1440/h, which even outperforms the k_{La} observed in typical benchtop fermentors. However, for microliter bioreactors, direct gas sparing is not a feasible method to control DO because of concerns over liquid evaporation. So oxygen supply in microliter bioreactors is usually achieved by membrane aeration. One option is to supply oxygen by *in situ* electrolytic gas generation through a thin gas-permeable membrane mounted on the bottom of the microreactor (35). In this case, large bubbles will form, which is not preferable for microliter reactors. Another option is to push the gas stream into a gas chamber with a gas-permeable membrane mounted on the top of the microreactor to allow diffusion of oxygen from the gas phase into the liquid phase without formation of gas bubbles (34,41–43). Under good mixing conditions, a k_{La} as high as 500/h can be achieved with comparable fermentation performance as that in benchtop fermentors.

5.5.4   pH

The most commonly used pH sensors for minibioreactor are optical sensors (optodes) based on fluorescence (41–43) and solid-state ion-selective field-effect transistor (ISFET) pH sensor chips (32,35). Optodes are cheap and easy to integrate into minibioreactors but have a relatively short lifetime and a narrow pH measurement range (pH 4–9). ISFET can operate at a wide pH range (pH 2–12) and high temperature (up to 120°C), but the ISFET sensor chips would need to be reused due to their high cost. Both types of sensors offer rapid and precise pH measurements over a long period of time.

While online monitoring of pH is reliable with commercially available pH sensors, pH control in minibioreactors is not well developed. The use of media with high pH buffering capacity is still the simplest way to control pH in minibioreactors. However, it is not always practical, especially when the cell density rises and the pH buffering capacity cannot compensate for the pH changes. Moreover, certain microorganisms cannot be cultivated in media that have high ion strength. Another option to control pH is by injecting acid or base intermittently into the minibioreactors (34,44). For milliliter bioreactors, adding a pH agent is not a major concern as long as the mixing is adequate. But for microliter reactors, pH control cannot be stable by this method. Another alternative approach to controlling pH is by dosing of CO_2 gas and NH_3 vapor (45). This approach has been implemented in commercial minibioreactors such as μ-24 reactors from Pall Corporation.

5.5.5   Cell Concentration

Due to volume limitations, it is impractical to take many samples from minibioreactors for cell concentration measurement. In minibioreactors, the cell concentration is normally measured in optical density (OD) by optical probes. Light from light-emitting diodes (LED) is guided into the reactor via optical fibers, passing through the vessel, and then received by a photodetector. Many minibioreactors have implemented this technology for OD measurement. For milliliter bioreactors, a microtiter plate reader can be used to measure OD at-line handled by a liquid handling robot (40).

OD measurement can only provide an estimate of total cell concentration without distinguishing viable cells from dead cells. Using impedance spectroscopy is an alternative way of measuring living cells by measuring the capacitance of cells, which is only detectable when the cell membrane is intact. This technology has been successfully integrated into a sensor array for a microbioreactor (32).

5.5.6 Feeding

Because many industrial fermentation processes are fed-batch fermentations, it is reasonable to evaluate strain performance and develop initial fermentation process in fed-batch mode. Various control strategies have been developed for benchtop fermentors as discussed in Reference (18). But feeding substrate in a controlled mode into a minibioreactor is a very challenging task due to the scale and the type of minibioreactors, especially for shaken minibioreactors.

One option to feed substrates in shaken minibioreactors is to use an enzymatic glucose auto-delivery system in which starch is digested by glucoamylase to release glucose (46). The commercial product of glucose auto-delivery system is available as EnBase, marketed by BioSilta (Oulu, Finland). This technology has been successfully used in shake flasks and microwells to achieve a relatively high cell density (OD_{600} > 30) and high protein expression in *E. coli*. It has been used to screen a protein expression library in fed-batch mode in microtiter plates and shake flasks to identify the best strain and to determine its optimal specific growth rate for scale-up in benchtop fermentors (47,48).

However, the glucose auto-delivery system has certain limitations for strain evaluation: (i) The amount of enzyme addition needs to be predetermined to ensure high cell growth rates under glucose-limited conditions. For high-throughput strain screening, this is not practical. (ii) Glucose is not always the carbon source of interest. In these cases, a more general method is preferable for substrate delivery. One example is to use a liquid handling system to deliver feed solution based on pH (40,49). For microliter bioreactors embedded on chips, it has been reported that a feeding system is embedded in the microreactor for a chemostat continuous culture (43).

5.6 COMMERCIAL HIGH-THROUGHPUT FERMENTATION SYSTEMS

While different types of minibioreactors have been reported, most of them are still in the prototype stage. Only a few have been commercialized and have the capability for high-throughput fermentation. For most research laboratories, it is preferable to consider using commercially available systems for strain evaluation and process development. In this section, the commercial high-throughput fermentation system is reviewed to provide guidelines for system selection.

5.6.1 Shaken Minibioreactors

Shaken minibioreactors are most widely used in high-throughput fermentation due to its high reliability, small footprint, rapid prototyping, and low cost. Use of microtiter plates as minibioreactors has been reviewed for their capability for oxygen transfer as a function of different well sizes and shapes (22). Integrating new DO and pH sensors into microtiter plates (50) and shake flasks (51) has greatly extended the applications of microtiter plates and shake flasks for strain evaluation and process development.

5.6.1.1 Disposable Shake Flask/Microtiter Plate with Integrated DO and pH Sensors Shake flasks are still the most widely used tool for initial strain screening and evaluation in majority of laboratories because of it is low-cost and easy to set up. By integrating DO and pH optical sensor spots into disposable shake flask as sold by PreSens (Regensburg, Germany), more information for pH and DO profiles during the culturing process can be obtained, thus making shake flasks economical high-throughput minibioreactors. These sensors are precalibrated and ready to use. The shake flasks are mounted on a shake flask reader that can monitor DO and pH online for up to nine shake flasks and log the data wirelessly into a PC (52). PreSens also offers 6-well, 24-well, and 96-well microwell plates integrated with DO or pH sensors called "SensorDishes." These disposable shake flasks and microwell plates offer a quick and cheap method of establishing simple high-throughput minibioreactor systems with reliable pH and DO online measurement. Scientific Industries (Bohemia, NY) also sells similar optical DO and pH sensors (named CellPhase) to be attached to any transparent flasks and vessels for real-time monitoring. The applications include media optimization and cell growth characterization, and so on.

5.6.1.2 Bioscreen C Bioscreen C was developed by the Finnish company Labsystems Oy (now Oy Growth Curves, Helsinki) in the mid-1980s initially for testing mutagenity/carcinogenity. It soon became a very useful tool for automating routine microbiology and was optimized for microbiology growth experiments. Bioscreen C is a computer-controlled incubator/reader/shaker that can run 200 samples with OD measurement within wave lengths ranging from 405to 600 nm. It is the first high-throughput microreactor with in-line turbidity measurement. Over the past 20 years, Bioscreen C has been used in many areas involving microbial growth curve measurement. Using Bioscreen C for media and growth condition optimization has been reported (53,54). However, the lack of measurement of other important fermentation parameters (e.g., pH and DO) has limited the application of Bioscreen C, which may finally be replaced by similar products but with more sensors integrated into the design.

5.6.1.3 BioLector The BioLector, from DASGIP AG (Jülich, Germany), is a system similar to Bioscreen but with optical pH and DO sensors integrated into 48-well and 96-well plates. Its optimized flower-shaped microwell can offer a k_{La} as high as 500/h. Because of the capability of online monitoring of pH, DO, optical cell density, and other products that can be measured by fluorescence such as green fluorescent protein (GFP) protein, BioLector is a powerful benchtop high-throughput fermentation system for strain screening, media optimization, and process development. Due to the use of standard format of microtiter plate, it can be integrated with liquid handling systems to add more capabilities such as pH control, nutrient feeding, and inducer addition (49).

5.6.1.4 μ-24 Bioreactor The μ-24 bioreactor, originally developed by Microreactor Technologies Inc (now a division of Pall Corporation), is a high-throughput fermentation system with real-time monitoring and control of temperature, pH, and DO. The diagram of each well with temperature, pH, and DO control can be found in Reference (55). The reactor cassettes are conforming to the SBS standard for 24-well culture plates with working volume of 1–7 mL. Each well can be controlled individually for temperature, pH, and DO with four control loops: temperature control loop by thermo conductor, DO control loop by air/oxygen aeration, two pH control loops with CO_2 as acid agent, and NH_3 as base agent. The detailed technologies for pH and DO sensors and controller are discussed in Reference (55). The μ-24 bioreactor cultivations for *Saccharomyces cerevisiae* demonstrated comparable growth to a 20-L stirred tank bioreactor fermentation in terms of offline metabolite and biomass analyses. High inter-well reproducibility was observed for process parameters such as online temperature, pH, and DO (45). A case study data on the use of μ-24 bioreactor for rapid process development can be found in Section 5.8.

5.6.2 Stirred Minibioreactor

The stirred minibioreactors use mechanic or magnetic stirring to provide mixing power and promote mass transfer rate. The oxygen transfer coefficient k_{La} in stirred minibioreactors can be close to that in benchtop fermentors, thus making the stirred minibioreactors a good replacement for benchtop fermentors in terms of mass transfer. In addition, the stirred minibioreactors can be used in applications where the viscosity of the medium is relatively high, such as high solid content fermentation. However, due to the cost of manufacturing, there are not that many commercial stirred minibioreactors on the market.

5.6.2.1 CellStation HTBR The CellStation HTBR from Fluorometrix (Stow, MA) is the first commercial high-throughput stirred minibioreactor. Detailed information on the product can be found in Reference (56). The system consists of 12 mini-stirred tank bioreactors (working volume of approximately 30 mL) equipped with disposable DO and pH optical sensing patches, a detector board, a gas distributor, and a turntable, which also serves as a water bath. All the bioreactors are positioned on the turntable, which is driven by a stepper motor underneath it. Agitation is provided by two 6 mm × 18 mm paddles powered by the agitation motors and can be adjusted continuously from 10 to 1000 rpm. Each bioreactor has a pH optode patch and a DO optode patch on the bottom for at-line measurement. Temperature is controlled by circulating water between a temperature-controlled circulator and the turntable. Gas mixtures sparging into each vessel are obtained by blending different gases through two flow meters. Process parameters in one bioreactor are measured one at a time (56). This system has been used in cell culture applications and the performance is comparable to that in disposable bag bioreactors (57). However, this product line was discontinued in late 2011 after Fluorometrix was acquired by Scientific Industries. A new version of high-throughput stirred minibioreactor system called BioGenie TriStation bioreactor has been developed by Scientific Industries.

5.6.2.2 *2mag Bioreactor 48* Another stirred minibioreactor launched by 2mag AG (Munich, Germany) is the 48-parallel milliliter bioreactor block originally developed by Professor Weuster-Botz at Technical University of Munich. Each baffled bioreactor has a nominal volume of 8–15 mL equipped with a gas-inducing impeller containing permanent magnet. The detailed view of one bioreactor with gas-inducing impeller mounted in the bioreactor block is depicted in Reference (58). The gas-inducing impeller rotates freely on a hollow shaft, which is mounted in the bioreactor block. Due to the rotation of the impeller, the medium is sucked in from the bottom and sterile gas is sucked in via the hollow shaft from the headspace of each bioreactor (40). The headspace is flushed continuously with sterile gas. Another type of impeller is also available for cultivation of mycelium-forming microorganisms as described in Reference (59). The bioreactor block fits a maximum of 48 stirred tank reactors, which are arranged in six rows each containing eight bioreactors. The bioreactor block is equipped with an electromagnetic multi drive, two heat exchangers (the first for temperature control of the bioreactors and the second for exhaust gas cooling) and a sterile gas supply. The gas distributor assures sterile gas distribution into all of the 48 stirred tank reactors and enables individual outlets for exhaust gas. The exhaust gas outlets at the same time serve as individual sampling ports. Optical fluorescent sensors for pH and DO are integrated in the bottom of each bioreactor, allowing online monitoring and control of these important state variables (44). The bioreactor block has been integrated into an automated experimental setup with a liquid handling system (40). The liquid handler can be used to automatically take samples as well as for realizing fed-batch processes and controlling pH individually for every single reactor. An additional microtiter

plate photometer allows for at-line analysis of OD, substrate and/or product concentrations, and so on. The process control software DASGIP fed-batch XP can store and present 3264 fermentation variables and control 144 set-point profiles.

A significant advantage of this high-throughput minibioreactors is the high oxygen transfer capability. The maximal k_{La} for the 48 stirred tanks with gas inducing impellers can be as high as 1400/h (40). The fermentation performance done in this 48-bioreactor system using different microorganisms such as *E. coli* (40), *Bacillus subtilis* (60,61), and *Streptomyces tendae* (59) is consistent with the data obtained from liter-scale benchtop fermentors.

5.6.3 Parallel Benchtop Fermentation System

As discussed in Sections 5.6.1 and 5.6.2, the advances in the technology of minibioreactor have made them popular in high-throughput fermentation development programs. But most commercially available minibioreactors still cannot replace the benchtop fermentation system in terms of reliability, flexibility, controllability, and scale-up/scale-down capability. Thus, commercial parallel benchtop fermentation systems can be another option used for high-throughput fermentation for process development with all the capabilities of a regular benchtop fermentor. In all commercial parallel benchtop fermentation systems, since each fermentor is independently controlled, each vessel can be used as a regular benchtop fermentor.

The BIOSTAT® Qplus from Sartorius is designed for parallel operation with high-throughput capability. The BIOSTAT® Qplus has the capability to control fully independently up to 12 culture vessels (0.5–1 L) with minimal manual operation. The system is equipped with powerful supervisory process control software for data acquisition, visualization, advanced process control, and recipe function.

Similar to BIOSTAT® Qplus, the DASGIP Parallel Bioreactor System is designed for parallel and controlled fermentation. Individual control of temperature, agitation, pH, and DO in up to 16 vessels in parallel allows DASGIP fermentation systems to accommodate most complex microbiology requirements. Flexible working volumes from 60 mL to 3 L provide a broad range of laboratory-scale experiment possibilities under aerobic and anaerobic conditions. By using the DASGIP Multipump Module for individual culture supplementation and substrate feed, batch and fed-batch operation is feasible. Using the DASGIP Gas Mixing Module, mass-flow-controlled individual blending of air, nitrogen, oxygen, and carbon dioxide for each vessel is precise and intuitive. The advantage of the DASGIP parallel bioreactor system is that most of the parts are modularized, which provides great flexibility for building customized parallel fermentation systems.

The GRETA multiple fermentation system, made by Belach Bioteknik AB (Stockholm, Sweden), was originally designed to produce recombinant proteins for structural genomics (62). The GRETA multifermenter system consists of bioreactor units, each encompassing six parallel fermentors. Up to four bioreactor units (24 parallel fermentors) can be integrated into one GRETA system. Each fermentor is built using stainless steel with either 0.5 or 1 L maximum working volume and equipped with a magnetic coupled bottom stirrer with combined impeller/sparger, sterile air filter, and an $O_2/N_2/$ air gas-mixing system. A single cleaning-in-place (CIP) station supports the 1–4 bioreactor units of a GRETA system. All parts in contact with media or product can be SIP. The system is also equipped with pneumatic syringes for automatic induction. OD is measured in each reactor by a built-in photometer. Calibration of sensors is performed in parallel and in a semi-automated mode to minimize setup time. Compared with other benchtop parallel fermentation systems, the GRETA system mostly resembles industrial SIP fermentation system with many automation features to minimize manual operation. This system is so far the best commercially available high-throughput fermentation system at liter scale for fermentation process development and optimization. The GRETA system has been used for DOE to develop and optimize the fermentation process for antibody production using *E. coli* (63).

5.7 TRENDS IN DEVELOPMENT OF HIGH THE GREATA-THROUGHPUT MINIBIOREACTOR SYSTEM

When developing or selecting a high-throughput fermentation system, it is better to integrate the fermentation system with other high-throughput unit operations such as strain screening, sample analysis, and product recovery. This kind of integrated high-throughput bioprocess can eliminate/minimize bottlenecks and improve productivity significantly (64). The microwell-type minibioreactors as discussed in Section 5.6.1 are a good fit for this framework. That is why currently most of commercial minibioreactors, except CellStation HTBR and 2mag bioreactor 48, are based on microwell-type reactors and can be easily integrated into the high-throughput framework by combining with liquid handling robots and microtiter plate readers (44,49).

By comparing the shaken and stirred minibioreactors, the stirred minibioreactors are closer to the benchtop fermentors in terms of mass transfer, mixing, process control, and sampling, while the shaken microbioreactors are more cost-effective. The 48-well multiple milliliter fermentors discussed in Section 5.6.2.2 is a comprehensive approach as it combined the advantages of shaken minibioreactors and stirred minibioreactors and has been applied for very complex fermentation processes such as filament fungi fermentation. If more feeding and control strategies can be implemented in this system, it will be almost indistinguishable from the regular benchtop fermentors.

On the other hand, in the case of microfluidic microbioreactor devices, the lack of cross-platform standardization and automation integration prevents them from being widely accepted by industrial bioprocess groups (64). So far all reports about microbioreactors based on microfluidics are laboratory prototypes. No commercial product is available. However, this type of mini-bioreactor can still find good application niches for fermentation process development. For example, the 150 μL microbioreactor reported in Reference (43) is equipped with a feed pump with pH, DO, and OD sensors embedded. This microbioreactor can be set up to perform a continuous chemostat culture, which cannot be implemented in current microwell minibioreactors. Chemostat cultivation of microorganisms offers unique opportunities for experimental manipulation of individual environmental parameter at a fixed, controllable specific growth rate. The chemostat method has been used as a tool to provide reproducible data for systematic physiology studies (65). However, it is labor-consuming to set up and maintain a chemostat in benchtop fermentors because of the long periods required to reach and maintain steady state after perturbation of a single parameter. The microbioreactor on the chip can be an excellent tool for performing chemostat continuous culture effectively in high-throughput formats to obtain intrinsic information regarding cell metabolism at various cultivation conditions.

5.8 CASE STUDIES OF FERMENTATION PROCESS DEVELOPMENT AND OPTIMIZATION USING HIGH-THROUGHPUT MINIBIOREACTORS

Due to the above-described advancements in the development of minibioreactors and parallel fermentation systems, now it is feasible for an industrial microbiology or biotechnology laboratory to choose the right combination of high-throughput fermentation systems for strain screening/evaluation, process development, and process scale-up/scale-down as illustrated in Figure 5.1B. For example, a set of minibioreactors can be used for strain screening, media optimization, and initial process development, and then the lead candidate strains can be moved forward for process development and optimization, scale-up/scale down studies in a set of parallel benchtop fermentation systems. This approach will accelerate process development activities in a high-throughput format while reducing the capital cost and increasing the probability of successful scale-up. In addition, using auto-substrate release technology described in Section 5.5.6, all the high-throughput screenings carried out in minibioreactors can be performed in fed-batch mode, which mimics the environment of high cell density culture more effectively (47).

Comparison of fermentation performance between minibioreactors and benchtop fermentors has been reported extensively (40,55,63,64,66,67). However, use of minibioreactors in the framework of high-throughput process development has not been reported frequently. Forty-eight parallel minibio-reactor systems have been used to establish a scale-down tool for riboflavin fed-batch fermentation using *B. subtilis* (61). Subsequently this minibioreactor system has been successfully applied to screen or discriminate four different riboflavin-producing *B. subtilis* strains. The results are validated by the data from liter-scale benchtop fermentors (60). Another example is reported on direct scale-up from a microwell minibioreactor to a 75-L pilot fermentor for *E. coli* recombinant protein expression. The scale-up is based on maintaining a similar level of oxygen transfer coefficient k_{La} at each scale. It was found that at high k_{La} level, the fermentation performance is consistent between the minibioreactor and the pilot fermentor (68).

5.8.1 Case Study 1: Protein Production

The ability to produce high-quality protein products in a timely and cost-efficient manner is of particular value throughout the discovery to commercialization stages of protein products useful in therapeutic, vaccine, and other industrial processes. To this effect, Pfenex Inc. (San Diego, CA) has developed a suite of toolboxes integrating strain, analytics, fermentation, and downstream process development for a *Pseudomonas fluorescens*-based expression platform. Seamless development of robust protein expression strains results from combining off-the-shelf toolboxes of expression plasmids that utilize a wide range of gene expression strategies with host strains of diverse phenotypes that are screened by high-throughput analytical methods. Subsequent simultaneous strain and far-reaching fermentation evaluation in mini and parallel bioreactor systems frequently results in high levels of soluble, properly folded protein production, which can be predictably scaled. These upstream platforms are coupled with high-throughput purification development involving rapid resin and process screens to ensure rapid development of production strain and process.

The fermentation evaluation process at Pfenex involves evaluating several strain candidates identified by the 96-well culture screen in multiple fermentation conditions in μ-24 minibioreactors in order to assess the potential of these strains to eventually perform in a scalable production process. Significant effort was made to develop a cultivation process in the minibioreactor to be predictable for a scalable high cell density fed-batch production process, with pH, temperature and DO monitoring and control. As demonstrated in this case study, the μ-24 minibioreactor has the capabilities to play a key role in expediting strain and fermentation development.

In this case study, the contributions of a mini and a parallel bioreactor system in the context of a fermentation development project are illustrated. A production strain isolate for a protein vaccine candidate selected at the 96-well-scale microtiter plate was evaluated under multiple fermentation conditions as part of a statistical design of experiments in the μ-24 minibioreactor. The various fermentations resulted in a 12-fold range of expression levels, with the identification of an optimum set of conditions, #2, for a further scale-up (Figure 5.4). Statistical model analysis of the data generated from this experiment highlights the effects of the different process variables and helps define the design spaces to predict the optimum ranges of the control variables

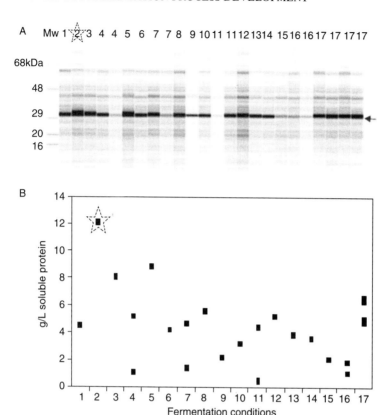

FIGURE 5.4. (A) Gel-like image of the sodium dodecyl sulfate–capillary electropho-resis (SDS-CGE) analysis of culture broth samples from the 24 fermentations using the 17 different fermentation conditions as part of a factorial experimental design. Molecular weight markers (Mw) are indicated by sizes in kDa. The numbers indicate the experimental number. The arrow points to the target protein expressed. (B) A plot of the concentrations of target protein expressed as determined by SDS-CGE in the 17 fermentation conditions tested. Repeat experiments are indicated by multiple dots. The star indicates fermentation with the best expression.

and interactions between them (Figure 5.5). The information generated is particularly useful during subsequent process development and further defini-tion of the ultimate production process.

The best fermentation conditions identified in the μ-24 bioreactor (#2) was subsequently confirmed to be scalable to the 20-L stainless steel conventional bioreactor scale in terms of cell growth, product titer, and productivity (Figure 5.6). The fact that high cell densities of greater than 200 OD can be obtained using the μ-24 bioreactor illustrates the superior oxygen and heat transfer capabilities of the minibioreactor system. In this particular case, it did require manual feeding of the carbon source to sustain the high cell density culture in the μ-24 bioreactor, but this operation should be automatable with liquid

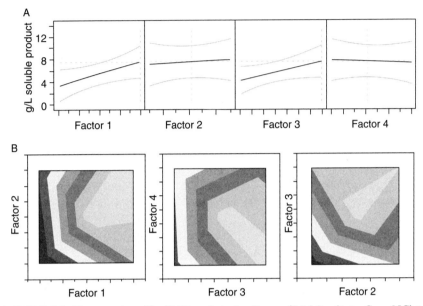

FIGURE 5.5. Plots produced by JMP statistical software (SAS Institute, Cary, NC) to highlight the main effects of the different fermentation parameters (factor) investigated (A) and to map the optimum ranges of the different factors and their interaction with each other (B).

handling systems. Table 5.2 illustrates how the process conditions identified in the μ-24 bioreactor enhanced expression level by approximately 15-fold over that in the 96-well culture. This enhancement was due in part to a 4.5-fold increase in cell density as a result of carbon feeding, better oxygen and pH control, and a 3.4-fold higher specific productivity as a result of a more optimum set of fermentation conditions. These contributions to the overall productivity were maintained during scale-up to the 20-L bioreactor.

5.8.2   Case Study 2: Antibody Fragment Expression

Several lead *Pseudomonas fluorescens* strains expressing an antibody fragment were identified to perform comparably to each other at the 96-well microtiter plate scale. When they were evaluated in multiple fermentations as part of a multivariate experimental design in the μ-24 bioreactor, overall expression levels were improved and differentiations between the strains became obvious. In a number of fermentations, the relative soluble titers of strains #2 and #4 were significantly higher than the remaining strains (Figure 5.7A). Nevertheless, as standard protocol, three lead strains in up to four lead fermentation conditions, a total of 12 fermentations were evaluated in a fully scalable high cell density fed-batch protocol in the 2-L parallel bioreactors.

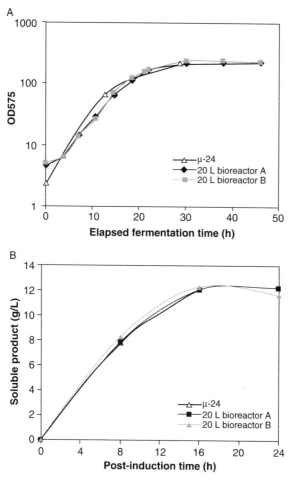

FIGURE 5.6. Growth curves (A) and productivity curves (B) of cultures grown in μ-24 minibioreactor and 20 L conventional stainless steel bioreactors to demonstrate scalability of the small scale process.

TABLE 5.2. Comparison of Cell Densities (OD), Volumetric Product Concentration, and Specific Productivities of Cultures Grown in a 96-Well Deep Well Microtiter Plate, μ-24 Minibioreactor, and 20-L Conventional Stainless Steel Bioreactors to Demonstrate Scalability

Fermentation Scale (Working Volume)	Final OD	Product Concentration (g/L)	Specific Productivity (mg/L/OD)
96-well (0.5 mL)	50	0.8	16
μ-24 (4 mL)	220	12	55
20 L bioreactor (10 L)	230	12	52

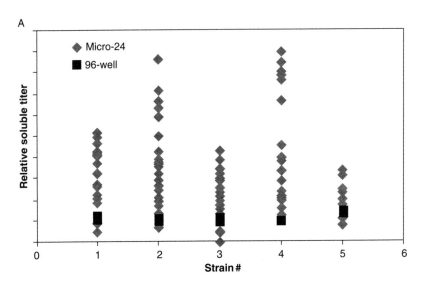

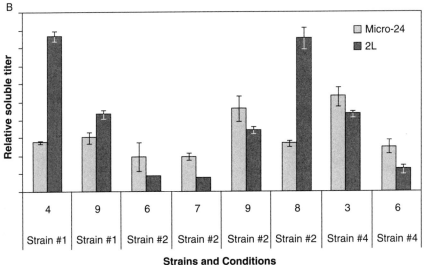

FIGURE 5.7. (A) Comparison of relative soluble product titers of cultures grown in a single set of fermentation conditions in a 96-well deep well microtiter plate versus those grown in multiple fermentation conditions in a μ-24 minibioreactor. (B) Relative soluble product titers of cultures grown in a μ-24 minibioreactor and 2-L conventional bioreactors to demonstrate the scalability of the small-scale process. The different fermentation conditions (3, 4, 6, 7, 8, and 9) and strains (#1, #2, and #4) are as specified.

In this evaluation, the scalability of six of the fermentations was confirmed at the 2-L scale (Figure 5.7B). However, two fermentations, strain #1 in fermentation condition #4 and strain #2 in fermentation condition #8, showed surprising increases in expression at the 2-L scale. The exact cause of this is not known, but a possible explanation is that some unknown fermentation parameter that had a significant effect on expression of this particular protein in these particular strains was controlled better at the 2 L. Nevertheless, strains #1 and #2 have subsequently been confirmed to be lead strains for expression of this particular antibody fragment in further studies.

These two case studies illustrate the range of results that one can encounter in the use of μ-24 bioreactors during simultaneous strain and fermentation evaluation. In one case, the scale-up of a strain and process from the μ-24 bioreactor to the 20-L scale bioreactor was successful and straightforward. In the second case, the scale-up was also successful but the results were not fully expected. These studies emphasize that with enough redundancies or contingencies built into the strain and fermentation development process, that is by evaluating more than one lead strain and one set of fermentation conditions, a high degree of success can be assured.

5.9 CONCLUSIONS AND THE PATH FORWARD

Current commercially available minibioreactors and parallel small-scale fermentation systems have enabled multidisciplinary metabolic engineering teams to quickly establish a fermentation research laboratory for high-throughput strain evaluation and rapid fermentation process development and optimization. Because of the small-scale and the high-throughput capability, the development cost and time can be reduced dramatically in a typical strain engineering program. Due to the availability of process monitoring and control capabilities, the DOE-type experiments for medium and growth condition optimization can be carried out under well-controlled and monitored conditions in minibioreactors. The parallel small-scale fermentors can be used to further validate fermentation performance, develop a scalable process, and challenge the process under scale-down conditions for identifying the operational windows for robustness. The advantage of this approach is that at the end of a scouting optimization program, a set of strains may be selected to fit different conditions and constraints at large scale. For example, depending on the limitation of OTR for different large-scale vessels, different strains instead of one can be picked up to give the best performance under the different OTR capacity of large-scale fermentors.

Furthermore, the mathematical model-based process development and optimization has a great advantage over traditional "trial and error" approaches. However, large amounts of reproducible historical fermentation data are needed to identify and validate the mathematical models before they can be used to predict and optimize the processes. The conventional high-throughput methods using regular microtiter plates or shake flasks can provide only end-point data, which is not sufficient to set up the models; thus, the models have to be identified using fully instrumented benchtop fermentors to generate dynamic process data sets. Nevertheless, using minibioreactors in strain evaluation, medium and growth condition optimization will generate massive reproducible dynamic data sets under well-controlled conditions for model identification and validation; thus, the model-based process optimization can now be implemented at an earlier stage of process development without depending on fermentations at benchtop scales. This enables evaluation of performance in benchtop fermentors at the optimal condition for each specific strain instead of using one standard set of conditions to evaluate all strains. Following characterization at the benchtop scale, the best candidate with its optimal fermentation conditions can be transferred for a further scale-up. As more and more sensors are integrated into minibioreactors, they can provide more dynamic process information to match the information quantity and quality typically obtained from benchtop fermentors, thus making it practical to develop strains and process more effectively.

REFERENCES

1. Warner, J.R., Reeder, P.J., Karimpour-Fard, A., Woodruff, L.B.A., and Gill, R.T. (2010) Rapid profiling of a microbial genome using mixtures of barcoded oligonucleotides. *Nat Biotechnol*, **28**, 856–862.

2. Gill, R.T., Wildt, S., Yang, Y.T., Ziesman, S., and Stephanopoulos, G. (2002) Genome-wide screening for trait conferring genes using DNA microarrays. *Proc Natl Acad Sci U S A*, **99**, 7033–7038.

3. Kennedy, M. and Krouse, D. (1999) Strategies for improving fermentation medium performance: a review. *J Ind Microbiol Biotechnol*, **23**, 456–475.

4. Bibhu, P.P., Ali, M., and Javed, S. (2007) Fermentation process optimization. *Res J Microbiol*, **2**, 201–208.

5. Guebel, D.V., Cordenons, A., Cascone, O., Giulietti, A.M., and Nudel, C. (1992) Influence of the nitrogen source on growth and ethanol production by *Pichia stipitis* NRRL Y-7124. *Biotechnol Lett*, **14**, 1193–1198.

6. Parrado, J., Millan, F., Hernandez-Pinzon, I., Bautista, J., and Machado, A. (1993) Sunflower peptones: use as nitrogen source for the formulation of fermentation media. *Process Biochem*, **28**, 109–113.

7. Weuster-Botz, D. and Wandrey, C. (1995) Medium optimization by genetic algorithm for continuous production of formate dehydrogenase. *Process Biochem*, **30**, 563–571.

8. Yee, L. and Blanch, H.W. (1993) Defined media optimization for growth of recombinant *Escherichia coli* X90. *Biotechnol Bioeng*, **41**, 221–230.

9. van der Werf, M.J., Jellema, R.H., and Hankemeier, T. (2005) Microbial metabolomics: replacing trial-and-error by the unbiased selection and ranking of targets. *J Ind Microbiol Biotechnol*, **32**, 234–252.

10. Bochner, B.R. (2003) New technologies to assess genotype–phenotype relationships. *Nat Rev Genet*, **4**, 309–314.

11. Bochner, B.R., Gadzinski, P., and Panomitros, E. (2001) Phenotype microarrays for high-throughput phenotypic testing and assay of gene function. *Genome Res*, **11**, 1246.

12. Bochner, B.R. (2009) Global phenotypic characterization of bacteria. *FEMS Microbiol Rev*, **33**, 191–205.

13. Carlson, R. and Carlson, J.E. (2005) Canonical Analysis of Response Surfaces: a valuable tool for process development. *Org Process Res Dev*, **9**, 321–330.

14. Liu, J., Xing, J., Chang, T., Ma, Z., and Liu, H. (2005) Optimization of nutritional conditions for nattokinase production by *Bacillus natto* NLSSE using statistical experimental methods. *Process Biochem*, **40**, 2757–2762.

15. Ratnam, B.V.V., Subba Rao, S., Mendu, D.R., Narasimha Rao, M., and Ayyanna, C. (2005) Optimization of medium constituents and fermentation conditions for the production of ethanol from *palmyra jaggery* using response surface methodology. *World J Microbiol Biotechnol*, **21**, 399–404.

16. Sun, J., Smets, I., Bernaerts, K., Van Impe, J., Vanderleyden, J., and Marchal, K. (2001) Quantitative analysis of bacterial gene expression by using the *gusA* reporter gene system. *Appl Environ Microbiol*, **67**, 3350.

17. Smets, I., Bernaerts, K., Sun, J., Marchal, K., Vanderleyden, J., and Van Impe, J. (2002) Sensitivity function-based model reduction: a bacterial gene expression case study. *Biotechnol Bioeng*, **80**, 195–200.

18. Lee, S.Y. (1996) High cell-density culture of *Escherichia coli*. *Trends Biotechnol*, **14**, 98–105.

19. Sun, J. (2012) Rapid strain evaluation using dynamic DO-stat fed-batch fermentation under scale-down conditions. In Cheng, Q. (ed.), *Microbial Metabolic Engineering, Methods in Molecular Biology*. SpringerLink, New York, vol. **834**, pp. 233–244.

20. Lara, A.R., Vazquez-Limón, C., Gosset, G., Bolívar, F., López-Munguía, A., and Ramírez, O.T. (2006) Engineering *Escherichia coli* to improve culture performance and reduce formation of by-products during recombinant protein production under transient intermittent anaerobic conditions. *Biotechnol Bioeng*, **94**, 1164–1175.

21. Neubauer, P. and Junne, S. (2010) Scale-down simulators for metabolic analysis of large-scale bioprocesses. *Curr Opin Biotechnol*, **21**, 114–121.

22. Schweder, T., Krüger, E., Xu, B., Jürgen, B., Blomsten, G., Enfors, S.O., and Hecker, M. (1999) Monitoring of genes that respond to process-related stress in large-scale bioprocesses. *Biotechnol Bioeng*, **65**, 151–159.

23. George, S., Larsson, G., and Enfors, S.O. (1993) A scale-down two-compartment reactor with controlled substrate oscillations: metabolic response of *Saccharomyces cerevisiae*. *Bioprocess Biosyst Eng*, **9**, 249–257.

24. Hewitt, C.J., Onyeaka, H., Lewis, G., Taylor, I.W., and Nienow, A.W. (2007) A comparison of high cell density fed-batch fermentations involving both induced and non-induced recombinant *Escherichia coli* under well-mixed small-scale and simulated poorly mixed large-scale conditions. *Biotechnol Bioeng*, **96**, 495–505.

25. Onyeaka, H., Nienow, A.W., and Hewitt, C.J. (2003) Further studies related to the scale-up of high cell density *Escherichia coli* fed-batch fermentations. *Biotechnol Bioeng*, **84**, 474–484.

26. Neubauer, P., Häggström, L., and Enfors, S.O. (1995) Influence of substrate oscillations on acetate formation and growth yield in *Escherichia coli* glucose limited fed-batch cultivations. *Biotechnol Bioeng*, **47**, 139–146.
27. Garcia-Ochoa, F. and Gomez, E. (2009) Bioreactor scale-up and oxygen transfer rate in microbial processes: an overview. *Biotechnol Adv*, **27**, 153–176.
28. Lacoursiere, A., Thompson, B.G., Kole, M.M., Ward, D., and Gerson, D.F. (1986) Effects of carbon dioxide concentration on anaerobic fermentations of *Escherichia coli*. *Appl Microbiol Biotechnol*, **23**, 404–406.
29. Ho, C.S. and Smith, M.D. (1986) Effect of dissolved carbon dioxide on penicillin fermentations: mycelial growth and penicillin production. *Biotechnol Bioeng*, **28**, 668–677.
30. McIntyre, M. and McNeil, B. (1997) Effects of elevated dissolved CO_2 levels on batch and continuous cultures of *Aspergillus niger* A60: an evaluation of experimental methods. *Appl Environ Microbiol*, **63**, 4171.
31. Shang, L., Jiang, M., Ryu, C.H., Chang, H.N., Cho, S.H., and Lee, J.W. (2003) Inhibitory effect of carbon dioxide on the fed-batch culture of *Ralstonia eutropha*: evaluation by CO_2 pulse injection and autogenous CO_2 methods. *Biotechnol Bioeng*, **83**, 312–320.
32. Krommenhoek, E.E., Gardeniers, J.G.E., Bomer, J.G., Li, X., Ottens, M., van Dedem, G.W.K., Van Leeuwen, M., van Gulik, W.M., van der Wielen, L.A.M., Heijnen, J.J., et al. (2007) Integrated electrochemical sensor array for on-line monitoring of yeast fermentations. *Anal Chem*, **79**, 5567–5573.
33. van Leeuwen, M., Krommenhoek, E.E., Heijnen, J.J., Gardeniers, H., van der Wielen, L.A.M., and van Gulik, W.M. (2009) Aerobic batch cultivation in micro bioreactor with integrated electrochemical sensor array. *Biotechnol Prog*, **26**, 293–300.
34. Lee, H.L.T., Boccazzi, P., Ram, R.J., and Sinskey, A.J. (2006) Microbioreactor arrays with integrated mixers and fluid injectors for high-throughput experimentation with pH and dissolved oxygen control. *Lab Chip*, **6**, 1229.
35. Maharbiz, M.M., Holtz, W.J., Howe, R.T., and Keasling, J.D. (2004) Microbioreactor arrays with parametric control for high-throughput experimentation. *Biotechnol Bioeng*, **85**, 376–381.
36. Kumar, S., Wittmann, C., and Heinzle, E. (2004) Review: minibioreactors. *Biotechnol Lett*, **26**, 1–10.
37. Gill, N.K., Appleton, M., Baganz, F., and Lye, G.J. (2008) Quantification of power consumption and oxygen transfer characteristics of a stirred miniature bioreactor for predictive fermentation scale-up. *Biotechnol Bioeng*, **100**, 1144–1155.
38. Kroneis, H.W. and Marsoner, H.J. (1983) A fluorescence-based sterilizable oxygen probe for use in bioreactors. *Sens Actuators*, **4**, 587–592.
39. Papkovsky, D.B. (1995) New oxygen sensors and their application to biosensing. *Sens Actuators B Chem*, **29**, 213–218.
40. Puskeiler, R., Kaufmann, K., and Weuster-Botz, D. (2005) Development, parallelization, and automation of a gas-inducing milliliter-scale bioreactor for high-throughput bioprocess design (HTBD). *Biotechnol Bioeng*, **89**, 512–523.
41. Szita, N., Boccazzi, P., Zhang, Z., Boyle, P., Sinskey, A.J., and Jensen, K.F. (2005) Development of a multiplexed microbioreactor system for high-throughput bioprocessing. *Lab Chip*, **5**, 819.

42. Zanzotto, A., Szita, N., Boccazzi, P., Lessard, P., Sinskey, A.J., and Jensen, K.F. (2004) Membrane-aerated microbioreactor for high-throughput bioprocessing. *Biotechnol Bioeng*, **87**, 243–254.

43. Zhang, Z., Boccazzi, P., Choi, H.-G., Perozziello, G., Sinskey, A.J., and Jensen, K.F. (2006) Microchemostat-microbial continuous culture in a polymer-based, instrumented microbioreactor. *Lab Chip*, **6**, 906.

44. Hortsch, R. and Weuster-Botz, D. (2010) Milliliter-scale stirred tank reactors for the cultivation of microorganisms. In Laskin, A.I., Sariaslani, S., and Gadd, G.M. (eds), *Advances in Applied Microbiology*. Elsevier, Burlington, MA, pp. 61–82.

45. Isett, K., George, H., Herber, W., and Amanullah, A. (2007) Twenty-four-well plate miniature bioreactor high-throughput system: assessment for microbial cultivations. *Biotechnol Bioeng*, **98**, 1017–1028.

46. Casteleijn, M.G., Neubauer, P., Panula-Perälä, J., Šiurkus, J., Vasala, A., and Wilmanowski, R. (2008) Enzyme controlled glucose auto-delivery for high cell density cultivations in microplates and shake flasks. *Microb Cell Fact*, **7**, 31.

47. Krause, M., Ukkonen, K., Haataja, T., Ruottinen, M., Glumoff, T., Neubauer, A., Neubauer, P., and Vasala, A. (2010) A novel fed-batch based cultivation method provides high cell-density and improves yield of soluble recombinant proteins in shaken cultures. *Microb Cell Fact*, **9**, 11.

48. Šiurkus, J., Panula-Perälä, J., Horn, U., Kraft, M., Rimšeliene, R., and Neubauer, P. (2010) Novel approach of high cell density recombinant bioprocess development: optimisation and scale-up from microlitre to pilot scales while maintaining the fed-batch cultivation mode of *E. coli* cultures. *Microb Cell Fact*, **9**, 35.

49. Huber, R., Ritter, D., Hering, T., Hillmer, A.K., Kensy, F., Müller, C., Wang, L., and Büchs, J. (2009) Robo-Lector—a novel platform for automated high-throughput cultivations in microtiter plates with high information content. *Microb Cell Fact*, **8**, 42.

50. John, G.T., Klimant, I., Wittmann, C., and Heinzle, E. (2003) Integrated optical sensing of dissolved oxygen in microtiter plates: a novel tool for microbial cultivation. *Biotechnol Bioeng*, **81**, 829–836.

51. Wittmann, C., Kim, H.M., John, G., and Heinzle, E. (2003) Characterization and application of an optical sensor for quantification of dissolved O2 in shake-flasks. *Biotechnol Lett*, **25**, 377–380.

52. Schneider, K., Schütz, V., John, G.T., and Heinzle, E. (2009) Optical device for parallel online measurement of dissolved oxygen and pH in shake flask cultures. *Bioprocess Biosyst Eng*, **33**, 541–547.

53. von Weymarn, N., Hujanen, M., and Leisola, M. (2002) Production of D-mannitol by heterofermentative lactic acid bacteria. *Process Biochem*, **37**, 1207–1213.

54. Zhang, J., Reddy, J., Buckland, B., and Greasham, R. (2003) Toward consistent and productive complex media for industrial fermentations: studies on yeast extract for a recombinant yeast fermentation process. *Biotechnol Bioeng*, **82**, 640–652.

55. Tang, Y.J., Laidlaw, D., Gani, K., and Keasling, J.D. (2006) Evaluation of the effects of various culture conditions on Cr(VI) reduction by *Shewanella oneidensis* MR-1 in a novel high-throughput mini-bioreactor. *Biotechnol Bioeng*, **95**, 176–184.

56. Ge, X., Hanson, M., Shen, H., Kostov, Y., Brorson, K.A., Frey, D.D., Moreira, A.R., and Rao, G. (2006) Validation of an optical sensor-based high-throughput bioreactor system for mammalian cell culture. *J Biotechnol*, **122**, 293–306.

57. Hanson, M.A., Brorson, K.A., Moreira, A.R., and Rao, G. (2009) Comparisons of optically monitored small-scale stirred tank vessels to optically controlled disposable bag bioreactors. *Microb Cell Fact*, **8**, 44.

58. Weuster-Botz, D., Puskeiler, R., Kusterer, A., Kaufmann, K., John, G.T., and Arnold, M. (2005) Methods and milliliter scale devices for high-throughput bioprocess design. *Bioprocess Biosyst Eng*, **28**, 109–119.

59. Hortsch, R., Stratmann, A., and Weuster-Botz, D. (2010) New milliliter-scale stirred tank bioreactors for the cultivation of mycelium forming microorganisms. *Biotechnol Bioeng*, **106**, 443–451.

60. Vester, A., Hans, M., Hohmann, H.-P., and Weuster-Botz, D. (2009) Discrimination of riboflavin producing *Bacillus subtilis* strains based on their fed-batch process performances on a millilitre scale. *Appl Microbiol Biotechnol*, **84**, 71–76.

61. Knorr, B., Schlieker, H., Hohmann, H.P., and Weuster-Botz, D. (2007) Scale-down and parallel operation of the riboflavin production process with *Bacillus subtilis*. *Biochem Eng J*, **33**, 263–274.

62. Hedrén, M., Ballagi, A., Mortsell, L., Rajkai, G., Stenmark, P., Sturesson, C., and Nordlund, P. (2006) GRETA, a new multifermenter system for structural genomics and process optimization. *Acta Crystallogr D Biol Crystallogr*, **62**, 1227–1231.

63. Larsson, J. (2005) High-throughput Fed-batch Production of Affibody® molecules in a novel Multi-fermentor system [Master Thesis]. Linköping University, Sweden.

64. Micheletti, M. and Lye, G. (2006) Microscale bioprocess optimisation. *Curr Opin Biotechnol*, **17**, 611–618.

65. Daran-Lapujade, P., Daran, J.-M., van Maris, A.J.A., de Winde, J.H., and Pronk, J.T. (2008) Chemostat-based micro-array analysis in Baker's yeast. In Poole, R.K. (ed.), *Advances in Microbial Physiology*. Academic Press, London, pp. 257–311, 414–417.

66. Kensy, F., Engelbrecht, C., and Büchs, J. (2009) Scale-up from microtiter plate to laboratory fermenter: evaluation by online monitoring techniques of growth and protein expression in *Escherichia coli* and *Hansenula polymorpha* fermentations. *Microb Cell Fact*, **8**, 68.

67. Betts, J.I., Doig, S.D., and Baganz, F. (2006) Characterization and application of a miniature 10 mL stirred-tank bioreactor, showing scale-down equivalence with a conventional 7 L reactor. *Biotechnol Prog*, **22**, 681–688.

68. Islam, R.S., Tisi, D., Levy, M.S., and Lye, G.J. (2008) Scale-up of *Escherichia coli* growth and recombinant protein expression conditions from microwell to laboratory and pilot scale based on matched k_{La}. *Biotechnol Bioeng*, **99**, 1128–1139.

6

THE CLAVULANIC ACID STRAIN IMPROVEMENT PROGRAM AT DSM ANTI-INFECTIVES

Bert Koekman and Marcus Hans

6.0 INTRODUCTION

Clavulanic acid is a naturally occurring antibiotic produced by *Streptomyces clavuligerus*. It has a weak antibiotic activity, but it is clinically important because it is a potent inhibitor of many β-lactamases. Due to this ability, it is used in combination with other β-lactam antibiotics to combat infections caused by β-lactamase-producing bacteria.

Its discovery, as a result from a screening program for β-lactamase inhibitors carried out at the Beecham Laboratories in England, was first reported in 1976 (1). One of the microbial cultures that gave a positive response in this screen was *S. clavuligerus* ATCC 27064 (*syn.* NRRL 3585), isolated from a South American soil sample (2), and the active component in the culture filtrate was shown to be a novel β-lactam, different from the other cephalosporins (cephamycin C) produced by *S. clavuligerus*.

To develop a commercially viable process, departing from a wild-type isolate producing only a small amount of therapeutically active substance requires substantial efforts in terms of optimization of cultivation conditions, up-scaling these conditions from shake flask format to large-scale stirred tank reactors, and modifying the genetic make-up of the strain to not only increase its production potential but also to improve other complex uncharacterized desirable fermentation phenotypes such as decreased viscosity and improved growth behavior. As will be set forth in the sections to follow, the interaction between fermentation and genetics disciplines is vital to successful process

Engineering Complex Phenotypes in Industrial Strains, First Edition. Edited by Ranjan Patnaik.
© 2013 John Wiley & Sons, Inc. Published 2013 by John Wiley & Sons, Inc.

development of strains bearing multiple complex traits, one of the challenges being the translation of large-scale process conditions to small-scale, high-throughput strain selection conditions and vice versa (see Chapter 5).

6.1 THE BIOSYNTHETIC PATHWAY TO CLAVULANIC ACID

Three gene clusters for clavulanic acid/clavam biosynthesis have been identified in *S. clavuligerus*: the clavulanic acid cluster, the clavam gene cluster, and a third cluster, the paralogous gene cluster, containing duplications from the first and second clusters. The biosynthetic pathway is shown in Figure 6.1, and

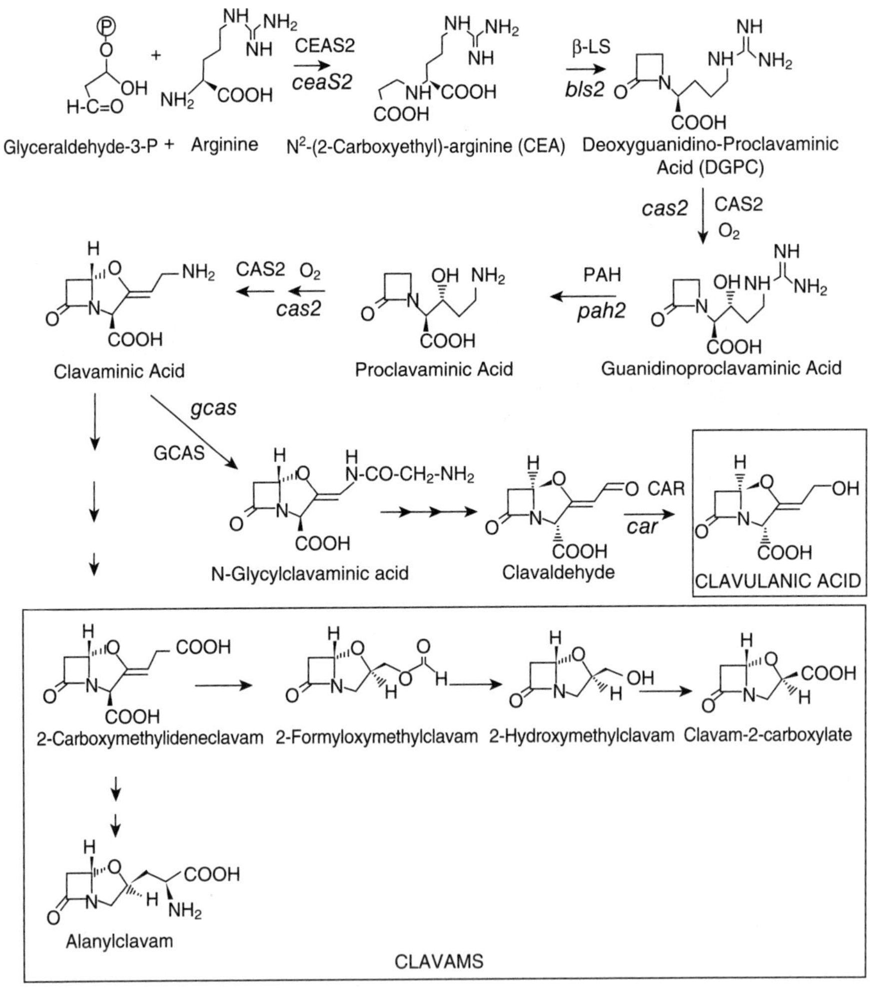

FIGURE 6.1. Biosynthetic pathway of clavulanic acid and clavams. Modified from Liras et al. (3), figure 3, with kind permission from Springer Science + Business Media.

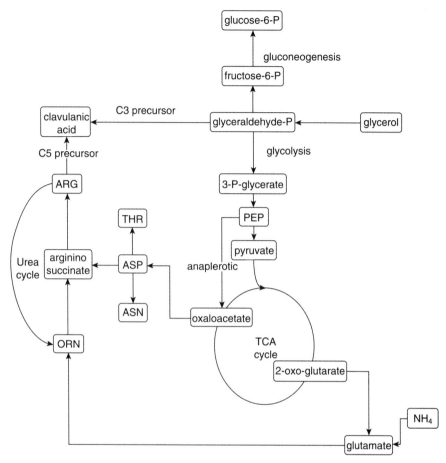

FIGURE 6.2. The position of clavulanic acid biosynthesis in central metabolism. After Bushell et al. (4). Reprinted from Enzyme Microb. Technol. 39, Michael E. Bushell, Samantha Kirk, Hong-Juan Zhao, Claudio Avignone-Rossa, Manipulation of the physiology of clavulanic acid biosynthesis with the aid of metabolic flux analysis, pp. 149–157 (2006), with permission from Elsevier. ARG, arginine; ORN, ornithine; THR, threonine; ASP, aspartic acid; ASN, asparagine; PEP, phosphoenolpyruvate; TCA, tricarboxylic acid.

the positioning of this pathway in the metabolic network providing the C5 precursor (arginine) and the C3 precursor (deriving from glycerol) is shown in Figure 6.2.

6.2 THE STRATEGY FOR IMPROVEMENT OF MULTIPLE COMPLEX PHENOTYPES

In 1992, Gist-brocades n.v., a fermentation industry based in Delft, the Netherlands, already having extensive activities in the production of β-lactam antibiotics, decided to develop its own production process for clavulanic acid. The strain development part of the program was ordered out on an exclusive basis with Panlabs, Inc. at its Taipei facility. Panlabs had already conducted preliminary strain improvement activities in 1991, starting with the culture collection strain, ATCC 27064, and was to resume this program for the first six years of the project (1993–1999).

At the time, classical strain improvement (CSI) (see Chapter 1) was the sole option to tackle the program, since knowledge on the pathway was scarce, biosynthetic genes were not cloned, and gene manipulation techniques were not available for *S. clavuligerus*. CSI does not require *a priori* knowledge referred to above, as it essentially introduces genetic variation in a random fashion by mutation, the outcome being determined by the selection applied.

In 1999, one year after the merging of Gist-brocades with DSM (Dutch State Mines, based in Heerlen, the Netherlands), the research agreement with Panlabs was discontinued, and further strain improvement was carried out in-house with DSM for the years to follow. The motivation underlying this decision was the closer interaction between the genetics and fermentation disciplines involved, which was deemed critical for the success of the project as it entered into a mature phase, necessitating scale-up of multiple complex phenotypes, often antagonistic to each other.

The last cycle of CSI was finalized in 2006, after which the focus of the project switched to genomics (see Chapter 3). In 2009, the company decided to divest its activities on the clavulanic acid market, and the research program was terminated.

6.3 RESULTS AND DISCUSSION

6.3.1 The Panlabs Years—Results from 1991 to 1999

In the initial years of the strain improvement program, the selection of superior mutants largely relied on so-called rational selections (see Chapter 1). Using this approach, a sub-population that is anticipated to be enriched for mutants with the phenotype of real interest, *viz.* improved productivity, is preselected from the mutant population to be used as input for the screen. The aim is to reduce the number of mutants to be examined by a few orders of magnitude compared with that of a brute-force random productivity screen. Of course, this will only produce a meaningful result when the rationale underlying the selection of the phenotype(s) in question is valid. For a review on the methodology, see, for example, References (5,6).

During the course of the strain improvement program, the following classes of rational selections were employed by Panlabs:

1. Selections aimed at increased product precursor availability: resistance to toxic analogues of arginine and related amino acids (ornithine, glutamic acid); resistance to inhibitors of glycerol uptake/utilization.

2. Selections aimed at relief of repression by C, N, or P: resistance to toxic analogues of glucose, ammonium, or phosphate; good clavulanic acid productivity in the presence of excess phosphate.

3. Detoxification of heavy metals that are sequestered by β-lactams: resistance to copper, nickel.

4. Resistance to the end product, clavulanic acid, itself.

5. Osmotolerance: resistance to high glycerol concentrations.

6. Resistance to agents not directly related to clavulanic acid biosynthesis, for example, protein synthesis inhibitors.

7. Selection of mutants with altered morphology on solid media.

8. Bioassay: selection of colonies producing large growth inhibition zones on a lawn of sensitive indicator bacteria (*Bacillus licheniformis*).

The specific selections that were used are listed in Table 6.1. The selections that were successful also appear in the Tables 6.2 and 6.4 of the strain lineage. Summarizing, all types of rationales, with the exception of (3), (4), and (7) have, at any time, yielded useful mutants during the course of the program.

Shortly after the start of the program, the strain lineage split into two branches, indicated by left and right (Figure 6.3). By far the most strains that have been in production derive from the right branch. Of the left branch, DS 30455 (culture #PF-19-41) was the sole strain to be used on production scale. However, because of its initially more favorable viscosity properties, this line was pursued for quite some time in parallel with the right branch, until it was abandoned in 2002 (results not shown in the table). In order to harvest the potential of the left branch after all (in the development of which a considerable amount of work had been invested for some years), an attempt was made to engineer its favorable properties into the right (production) branch. However, efforts to merge both lines by protoplast fusion (see Chapters 1 and 4) remained unsuccessful. The inability to easily migrate a complex phenotype from one strain lineage to another is a real limitation for the widespread applicability of CSI in engineering complex phenotypes.

An overview of the strain lineage is shown in Figure 6.3. A detailed list of the strains selected until the strain improvement program moved to DSM is shown in Table 6.2. During the course of the strain improvement program, various mutagens were used to induce genetic variation (e.g., ultraviolet irradiation, alkylating agents, and nitrous acid). However, from the table of the strain lineage it becomes apparent that virtually all improved mutants were selected from populations obtained by mutagenesis with alkylating agents (mostly NTG, occasionally EMS, see Chapter 1), that is, mutagens independent of the SOS-repair pathway. It is suspected that this pathway might be absent in *S. clavuligerus*.

The product titer development of the right branch in shake flask from the start of the program until 1996 is shown in the graph in Figure 6.4. The graph shows that productivity increases tend to become smaller with the progression

TABLE 6.1. Selective Agents Employed in Rational Selections

Class 1A—C5 precursor availability	
α-amino-δ-hydroxyvaleric acid	Ornithine analogue
Arginine hydroxamate	Arginine analogue
Canaline	Ornithine analogue
Canavanine	Arginine analogue
Homo-arginine	Arginine analogue
(DL)-lysine	Arginine analogue
Methionine sulfoxide	Glutamate (ornithine precursor) analogue
Methionine sulfoximine	Glutamate (ornithine precursor) analogue
α-methylornithine	Ornithine analogue
Class 1B—C3 precursor availability	
L-cysteine	Glycerol transport inhibitor
Ethanolamine	Glycerol transport inhibitor
β-mercaptoethanol	Glycerol transport inhibitor
Methylglyoxal	Glycerol utilization inhibitor
L-serine	Glycerol transport inhibitor
Class 2—Relief of catabolite repression	
2-deoxyglucose	Glucose analogue
Methylamine	Ammonium analogue
Tetramethylammoniumchloride	Ammonium analogue
Trimethylamine	Ammonium analogue
Sodium arsenate	Phosphate analogue
Sodium vanadate	Phosphate analogue
Class 6—Antibiotic resistances	
Chloramphenicol	
Erythromycin	
Fusidic acid	
Kasugamycin	
Streptomycin	
Tetracycline	

of the program, as is commonly experienced in classical strain improvement programs. The selection conditions used in shake flasks were established by down-scaling the production conditions in so far as possible. Thus, in the first years of the program, selections were carried out in complex media, derived from the recipe used on production scale. Many different variants have been in use, due to frequent rebalancing of the recipe in order to make the most of the potential of newly selected strains, but the important medium components shared in common are soy flour and casein hydrolysate as nitrogen sources, and glycerol and sometimes maltodextrins as carbon source, as *S. clavuligerus* is unable to grow on glucose. Initially, phosphate was used as buffer, but this was largely replaced by N-morpholino-propanesulfonic acid (MOPS) to prevent phosphate repression. During the course of the program, there has

TABLE 6.2. Strains Selected at Panlabs

Year	Culture#	Parent	Mutagen	Selection	Production#
1991	1991-10-125	ATCC 27064	—	cnv^R (1)	—
	1991-41-145	1991-10-125	NTG	Random	—
1993	PF-19-2	1991-41-145	—	Re-isolation	—
	23-82	PF-19-2	NTG	mont^R (1)	—
	PF-19-5	23-82	—	Re-isolation	—
	PF-19-24	PF-19-5	NTG	mont^R (1)	—
1994	PF-19-41	PF-19-24	NTG	cml^R (6)	DS30455
1993	PF-19-10	PF-19-2	NTG	mont^R (1)	—
	31-150R3	PF-19-10	—	Re-isolation	—
1994	PF-19-35	31-150R3	NTG	PO$_4$ (2)	—
	PF-19-85	PF-19-35	EMS+UV	Random	—
	PF-19-130	PF-19-85	NTG	Bioassay (8)	—
1995	PF-19-188	PF-19-130	NTG	Bioassay (8)	—
	PF-19-230	PF-19-188	NTG	Bioassay (8)	DS31810
1996	PF-19-307	PF-19-230	—	Re-isolation	DS33037
1997	PF-19-381	PF-19-307	NTG	Osmotol. (5)	—
	PF-19-409	PF-19-381	NTG	cml^R (6)	—
1998	PF-19-429	PF-19-409	NTG	Osmotol. (5)	—
	PF-19-432	PF-19-409	NTG	Random	DS33871
1999	PF-19-440	PF-19-429	NTG	Osmotol. (5)	—
	PF-19-446	PF-19-440	NTG	PO$_4$ (2)	—
	PF-19-450	PF-19-446	NTG	Defin^d. med.	DS36063

Year refers to the first isolation of the culture. When a strain was adopted as a production strain (entry in column **Production#**), there is usually a considerable lapse of time (~0.5–1 year) after its first isolation.
Selection: the type of rationale (see list in Section 6.3.1) is stated in brackets. cml^R, chloramphenicol resistance; cnv^R, canavanine (arg analogue) resistance; mont^R, α-methylornithine (ornithine analogue) resistance; random, no (rational) preselection prior to shake flask testing.
Mutagen: NTG, N-methyl-N'-nitro-N-nitrosoguanidine; EMS + UV: ethyl methane sulfonate assisted by ultraviolet irradiation.
The first seven entries represent the left branch; the other entries represent the right branch.

been a tendency toward lighter recipe strengths to prevent oxygen limitation during the fermentation and to better reflect the intrinsic production potential of the mutants. In Table 6.3, representative examples of media are given.

From 1997 onward, (semi-)defined media were developed that lacked the particulate component, soy flour. Although the titers obtained in these media were lower than in the complex ones in absolute sense, productivities had advanced sufficiently during the improvement program to afford this lapse, and still to enable discrimination of superior cultures. In the end, it was felt that development toward "cleaner" recipes would be more advantageous.

The last strain to be delivered by Panlabs (PF-19-450, to become production strain DS36063) was selected on semi-defined medium in 1999. Thereafter, the

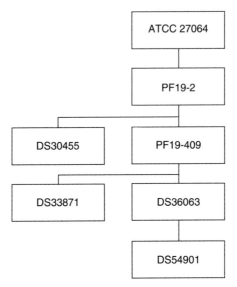

FIGURE 6.3. A bird's eye overview of DSM's clavulanic acid lineage. For further details, see the tables in Sections 6.1 and 6.2.

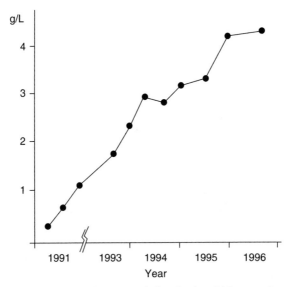

FIGURE 6.4. Productivity development of clavulanic acid in complex media in shake flasks.

TABLE 6.3. Media Compositions

Solid Media

Ingredients (g/L)	Plate Agar	Sporulation Agar
Glucose. aq.	10	10
Casein hydrolysate	2	2
Yeast extract	1	1
Beef extract	1	1
$CaCO_3$	1	5
Trace elements #1 (mL)	5	5
Bacto agar	20	10
Oatmeal agar	—	30

Trace Elements

Ingredients (g/L)	Cocktail #1	Cocktail #2
H_2SO_4 concentrated	—	20.4
Citric acid. aq.	—	50
$Fe(NH_4)_2HC_6H_5O_7$	2.7	—
$ZnSO_4$. 7aq.	2.8	16.75
$CuSO_4$. 5aq.	0.125	1.6
$CoCl_2$. 6aq.	0.1	—
$MnSO_4$. aq.	1.2	—
$MnCl_2$. 4aq.	—	1.5
$Na_2B_4O_7$. 10aq.	0.16	—
H_3BO_3	—	2
Na_2MoO_4. 2aq.	0.054	2

Production Media

Ingredients (g/L)	CM-3	50% CM-3	SM-2	50% MM-1
Casein hydrolysate	15	7.5	5	1.25
Soy flour	15	7.5	—	—
Asparagine	—	—	3.5	0.9
Glycerol	50	25	30	7.5
MOPS	10	10	10	10
KH_2PO_4	1.35	0.7	0.7	0.2
$MgSO_4$. 7aq.	0.6	0.3	0.3	0.1
$CaCl_2$. 2aq.	0.35	0.2	0.2	0.05
$FeSO_4$. 7aq.	0.45	0.2	0.2	0.05
Trace elements #2 (mL)	1.9	1	1	0.25
Basildon (mL)	0.2	0.2	0.1	—

Solid media: Presterile pH 6.7, no further adjustment after autoclaving (20 minutes. 120°C).
Production media: Presterile pH 7.0, no further adjustment after autoclaving (20 minutes. 120°C). CM-3 and 50% CM-3 are complex recipes (standard and light variant), SM-2 and 50% MM-1 are semi-defined (the latter was especially designed for use in microtiter formats).
Conditions for growth and production: 26.5–28°C, orbital shaker at 280 rpm, harvest time 96–120 hours.

program was transferred to DSM, where semi-defined media became routinely included in the selection of mutants. In the declining years of the program, the fermentation process was adapted by replacing the most expensive medium ingredient, casein hydrolysate, by cheaper nitrogen sources (wheat gluten, pea protein), and the media used for strain selection were changed accordingly by small-scale optimization, to mimic the new large-scale conditions as far as possible, just as these conditions were established at the start of the program (see above). However, in contrast to the situation at that time, a set of consecutive production strains with ascending titers had become available by now, which was helpful in validating shake flask protocols in terms of reflecting the correct order of these strains.

6.3.2   The DSM Years—Results from 1999 to 2006

When DSM proceeded with the strain improvement program, the efforts shifted from rational preselections to high-throughput miniaturized prescreening in 96-well microtiter plate formats (see Chapter 5). Two conditions had to be met to make this type of screening possible:

1. The development of semi-defined media without insoluble components (see previous section and Table 6.2—the light recipe MM-1 was especially designed for this purpose).
2. The development of high-throughput analytical techniques. This was realized by replacing the high-performance liquid chromatography (HPLC) assay method used with Panlabs by stopped-flow nuclear magnetic resonance (NMR) analysis capable of handling large numbers of samples.

In the first three cycles of mutation and selection, the number of mutants processed by the MTP screen was still limited, but from 2003 onward, screening capacity reached throughputs of approximately 60,000 (selections with DS41411 as the progenitor) to 110,000 (with DS48802). A list of the strains selected at DSM is shown in Table 6.4; see also the overview of the lineage in Figure 6.3.

TABLE 6.4. Strains Selected at DSM

Year	Strain	Parent	Mutagen	Selection
2000	DS37853	DS36063	NTG	MTP
2001	DS39876	DS37853	EMS	Random
2002	DS41411	DS39876	NTG	str^R (6)
2003	DS48802	DS41411	NTG	MTP
2005–2006	DS54901	DS48802	nUV+8-MOP	MTP

See Table 6.2; str^R, streptomycin resistance; nUV+8-MOP, near UV (365 nm) as sensitizer in combination with the cross-linking agent, 8-methoxypsoralen.

In contrast with the strains selected at Panlabs, of which only a limited number attained the production status, as indicated by the DS number in the last column, all strains listed in Table 6.4 have been production strains. The reason for this is the closer interaction between genetics and fermentation groups as both disciplines operated on the same site since the transfer of the program in 1999: the decision to enter into a next cycle of mutation and selection was made only after the new progenitor strain had been successfully introduced on production scale. Thus, selections operated in a campaign-wise fashion rather than being conducted continuously. There is a hiatus in Table 6.4 in 2004 due to an earlier campaign with strain DS48802, which failed to yield an improved strain. For this reason, the campaign was repeated a year later, using an alternative mutagen, which resulted in the selection of strain DS54901. Again, all progenitors of consecutive cycles of mutation and selection are mutants induced by SOS-independent mutagens (see previous section). DS54901 was the last strain to be used in production, from 2007 until the close-down of the plant at the end of 2009.

The above example reiterates the notion that scale-up of complex traits such as production titers requires not only integration between the high-throughput strain evaluation assays and the fermentation group but also the ability to continuously validate and improve assay conditions to mimic production fermentation.

6.4 FUTURE PERSPECTIVES

Recent insights into *S. clavuligerus* genomics have generated new options for targeted engineering approaches. The biosynthetic gene cluster of clavulanic acid is located contiguous to the cephamycin gene cluster in the genome of *S. clavuligerus* (7). The region is ~15 kb in size and includes all genes of the biosynthetic enzymes needed for clavulanic acid formation. Starting from L-arginine, clavulanic acid is formed via a series of at least seven enzymatic steps. Key enzymes are β-lactam synthase (BLS) and clavaminate synthase (CAS). Noteworthy, a regulatory protein, ClaR, has been identified as well. It was shown earlier that this protein is involved in the regulation of the late steps of clavulanic acid biosynthesis (8).

While those findings shed light on the mechanisms of clavulanic acid formation, it became clear that elucidation of the genome sequence would prove valuable for further targeted engineering approaches. With the aid of bioinformatics, metabolic network knowledge combined with transcript analysis data (see Chapter 3) will provide leads to improving the yield of clavulanic acid production. Leads can include the overproduction of biosynthetic enzymes, fine-tuning of regulator abundance or activity, formation and transport of precursors, but also pathways that influence the morphology of the cells.

In the last years, progress has been made to elucidate the genome sequence of *S. clavuligerus*. Further system biological comparisons of wild-type and classically improved strains revealed important insights that can be used for future rational strain development programs (9,10). This chapter highlights the findings of those two publications.

Genome sequencing of the wild-type strain *S. clavuligerus* ATCC 27064 revealed several remarkable features of this member of the *Streptomyces* family. On the one hand, the observed genome size of 6.7 Mb is quite small compared with that of other sequenced *Streptomyces* bacteria. Interestingly, a giant linear plasmid that has a length of 1.8 Mb was identified. Although earlier reports suggested the presence of more than one plasmid, in the reported sequencing project only two replicons were identified, indicative of the presence of only one plasmid. Furthermore, the sum of the putative protein encoding genes on those two replicons matched well with the typical makeup seen in other *Streptomyces* genomes. Therefore, it can be seen as certain that *S. clavuligerus* indeed possesses one large linear plasmid.

Analysis of the 1.8-Mb plasmid revealed several interesting features. The replicon contained no genes crucial for the metabolism of *S. clavuligerus*. All stable RNAs necessary for primary metabolism (rRNA and tRNA) are encoded on the main chromosome. Therefore, the megaplasmid seems to be dispensable for the core metabolism of *S. clavuligerus*. Strikingly, the plasmid is packed with secondary metabolite gene clusters. No fewer than 25 such gene clusters, a number of the same order as observed in the chromosomes of other *Streptomyces* genomes, were dispersed throughout the plasmid. Together with the clusters identified on the chromosome, the total number of putative secondary metabolite gene clusters identified in *S. clavuligerus* is 48. As expected, the three known antibiotic gene clusters were identified in the genome assembly. While the supercluster encoding the clavulanic acid and cephamycin C biosynthetic pathways (SMC10-11) and one of the clavam clusters (SMC9) are on the main chromosome, the alanylclavam cluster (SMCp13) is located on the megaplasmid. There are indications that cross-regulation takes place between the megaplasmid and the chromosome. A gene encoding a γ-butyrolactone receptor protein (ScaR/Brp) was identified and shown to regulate clavulanic and cephamycin C production. It turned out that the only copy of the *brp* gene is located on the megaplasmid, which is remarkable, because all other characterized γ-butyrolactone receptors are located on the chromosome. Moreover, it also means that Brp transregulates several factors on the chromosome (at least the clavulanic acid and cephamycin C gene clusters).

Looking at the possible evolution of the megaplasmid in *S. clavuligerus*, the central position of the origin of replication suggested that multiple recombination events within the chromosome took place that finally led to megaplasmid pSCL4; this theory is in contrast to the situation in *Streptomyces coelicolor* A3(2), where likely one single crossover between a 365-kb plasmid and the chromosome led to the formation of a 1.8-Mb plasmid, a plasmid very similar to the *S. clavuligerus* pSCL4.

After gaining insight in the genome organization of *S. clavuligerus*, the authors chose a functional genomics approach (see Chapter 3) to elucidate transcription and expression differences between wild-type and classically improved industrial *S. clavuligerus* strains. By doing so, more could be learned about metabolic changes induced by strain improvement, and ultimately targets and hints would be obtained for future strain improvement programs.

In order to better understand and predict the metabolism, the metabolic fluxes during antibiotic production were computationally predicted, using a constraints-based genome-scale metabolic network model of *S. clavuligerus* (10). By using Affymetrix microarray gene chips (Affymetrix, Cleveland, OH), the transcript levels of the wild-type strain ATCC 27064 and industrial strain DS48802 during the stationary phase were elucidated and compared.

First, it turned out that almost all genes associated in the clavulanic acid gene cluster were overexpressed significantly (between two- and eightfold) in the industrial strain compared with the wild-type strain. Interestingly, the pathway specific regulator genes *claR* and *ccaR* are also overexpressed in the industrial strain. They are located within the same supercluster and their products have been shown to regulate clavulanic acid production positively (11). Importantly, hybridization of *S. clavuligerus* DS48802 genomic DNA revealed no amplifications of genes of the clavulanic acid cluster, as observed for other industrial strains such as the industrial kanamycin producers *Streptomyces kanamyceticus* (12). Therefore, the overexpression observed appeared to be caused by transcriptional (and post-translational) changes only.

Second, the obtained transcriptomic data correlate well with the flux balance analysis of increased clavulanic acid production. For this test, a constraints-based genome-scale metabolic model of *S. clavuligerus* was developed and the flux changes during increased production of clavulanic acid were dynamically modeled. Forty percent of the genes that showed increased transcript levels (fold change >2) were also predicted to do so using the described metabolic network model. Although 40% might appear to be not a large percentage, it is still statistically very significant. The observed increase in the clavulanic acid cluster gene expression seems to be a crucial change for antibiotic overproduction in this strain. A complete redirection of primary metabolism seems not to be necessary for overproduction.

Furthermore, some significant changes in primary metabolism gene expression could still be observed and correlated to clavulanic acid overproduction. Glycerol uptake and metabolism is clearly upregulated, indicating improved utilization of glycerol as a carbon source and increased production of the clavulanic acid precursor, G3P. Moreover, aconitase and citrate synthase from the citric acid cycle are downregulated, which is likely to result in an increased intracellular G3P pool. This situation is remarkably similar to the result of the rationally designed *gap1* deletion (13) that blocked G3P conversion into 1,3-diphosphoglycerate, thus improving clavulanic acid biosynthesis by increasing the intracellular G3P pool. Also, a significant upregulation of glutamine synthetases and glutamate importers have been observed. Glutamate can serve as a source for biosynthesis of the clavulanic acid precursor arginine.

To conclude, the data generated by Medema et al. showed that a strain improvement program by random mutagenesis and screening caused gene transcript changes in primary as well as secondary metabolism. The overlap with results obtained by rational metabolic engineering is intriguing. New leads from transcript changes in those studies, such as the increased transcription of glutamine and glutamate synthetase, and those encoding several transporters, can be combined to rationally design novel clavulanic acid high producer strains.

ACKNOWLEDGMENTS

The authors gratefully acknowledge the invaluable contribution of Anderson Hong, Anne Hsieh, and coworkers at Panlabs, where the strain improvement program was conducted during its first years.

REFERENCES

1. Brown, A.G., Butterworth, D., Cole, M., Hanscomb, G., Hood, J.D., Reading, C., and Rolinson, G.N. (1976) Naturally occurring β-lactamase-inhibitors with antibacterial activity. *J Antibiot*, **29**, 668–669.
2. Higgens, C.E. and Kastner, R.E. (1971) *Streptomyces clavuligerus* sp. nov., a β-lactam antibiotic producer. *Int J Syst Bacteriol*, **21**, 326–331.
3. Liras, P., Gomez-Escribano, J.P., and Santamarta, I. (2008) Regulatory mechanisms controlling antibiotic production in *Streptomyces clavuligerus*. *J Ind Microbiol Biotechnol*, **35**, 667–676.
4. Bushell, M.E., Kirk, S., Zhao, H.-J., and Avignone-Rossa, C.A. (2006) Manipulation of the physiology of clavulanic acid biosynthesis with the aid of metabolic flux analysis. *Enzyme Microb Technol*, **39**, 149–157.
5. Rowlands, R.T. (1984) Industrial strain improvement: mutagenesis and random screening procedures. *Enzyme Microb Technol*, **6**, 3–10.
6. Rowlands, R.T. (1992) Strain improvement and strain stability. In Finkelstein, D.B. and Ball, C. (eds.), *Biotechnology of Filamentous Fungi*. Butterworth-Heinemann, Boston, pp. 41–64.
7. Mellado, E., Lorenzana, L.M., Rodríguez-Sáiz, M., Díez, B., Liras, P., and Barredo, J.L. (2002) The clavulanic acid biosynthetic cluster of *Streptomyces clavuligerus*: genetic organization of the region upstream of the *car* gene. *Microbiology*, **148**, 1427–1438.
8. Paradkar, A.S., Aidoo, K.A., and Jensen, S.E. (1998) A pathway-specific transcriptional activator regulates late steps of clavulanic acid biosynthesis in *Streptomyces clavuligerus*. *Mol Microbiol*, **27**, 831–843.
9. Medema, M.H., Trefzer, A., Kovalchuk, A., Van Den Berg, M.A., Müller, U., Heijne, W.H.M., Wu, L., Alam, M.T., Ronning, C.M., Nierman, W.C., Bovenberg, R.A.L., Breitling, R., and Takano, E. (2010) The sequence of a 1.8-Mb bacterial linear

plasmid reveals a rich evolutionary reservoir of secondary metabolic pathways. *Genome Biol Evol*, **2**, 212–224.

10. Medema, M.H., Alam, M.T., Heijne, W.H.M., Van Den Berg, M.A., Müller, U., Trefzer, A., Bovenberg, R.A.L., Breitling, R., and Takano, E. (2011) Genome-wide expression changes in an industrial clavulanic acid overproduction strain of *Streptomyces clavuligerus*. *Microb Biotechnol*, **4**, 300–305.

11. Alexander, D.C. and Jensen, S.E. (1998) Investigation of the *Streptomyces clavuligerus* cephamycin C gene cluster and its regulation by the CcaR protein. *J Bacteriol*, **180**, 4068–4079.

12. Yanai, K., Murakami, T., and Bibb, M. (2006) Amplification of the entire kanamycin biosynthetic cluster during empirical strain improvement of *Streptomyces kanamyceticus*. *Proc Natl Acad Sci U S A*, **103**, 9661–9666.

13. Li, R. and Townsend, C.A. (2006) Rational strain improvement for enhanced clavulanic acid production by genetic engineering of the glycolytic pathway in *Streptomyces clavuligerus*. *Metab Eng*, **8**, 240–252.

7

METABOLIC ENGINEERING OF RECOMBINANT *E. COLI* FOR THE PRODUCTION OF 3-HYDROXYPROPIONATE

Tanya Warnecke Lipscomb, Matthew L. Lipscomb, Ryan T. Gill, and Michael D. Lynch

7.0 INTRODUCTION TO BIOSYNTHESIS OF 3-HYDROXYPROPIONIC ACID

Bioprocesses directed toward the production of commodity chemicals from renewable resources have become a major focus of the chemical industry. Organic acids are one group that comprises a significant portion of the proposed commodity chemical market from bioprocesses. One particular organic acid, 3-hydroxypropionate (3-HP), has been identified as a highly attractive potential chemical feedstock for the production of numerous large market commodity chemicals that are currently derived from petroleum (1). Commodity products that can be readily produced using 3-HP include acrylic acid, 1,3-propanediol, methyl-acrylate, and acrylamide. The current estimated global market value of acrylic acid alone exceeds $10 billion annually.

Research in the biotechnology arena during the past decade has encompassed several key categories, including the efficient extraction of carbon sources from waste biomass, overall biocatalyst development, and downstream processes required for commercialization of valuable products. The field of biocatalyst development has grown from the humble beginning of overexpression of native metabolites to the rational, holistic microbial engineering design

Engineering Complex Phenotypes in Industrial Strains, First Edition. Edited by Ranjan Patnaik.
© 2013 John Wiley & Sons, Inc. Published 2013 by John Wiley & Sons, Inc.

for production of heterologous compounds. This case study will focus primarily on biocatalyst development through understanding organic acid tolerance and the engineering of a tolerant strain for the economical fermentative production of 3-HP from a renewable feedstock. 3-HP is a naturally occurring metabolite produced at low levels by several photosynthetic microorganisms that use a carbon fixation cycle termed the 3-HP cycle (2–4). However, for all microorganisms tested to date in our laboratory, including various gram-negative and gram-positive microorganisms, 3-HP displays toxicity at levels as low as 20 g/L in minimal media. This observed inhibition, which occurs by shutting down various metabolic pathways at titer levels well below what are needed for successful commercialization, will be discussed in detail below.

7.1 ORGANIC ACID TOXICITY

Evolving tolerance to organic acids requires approaches reliant on understanding of how the acid affects the organism as a whole. Traditional selection studies (see Chapter 1) remain focused on the identification of a single genetic element or mutant associated with a measureable improvement in a given phenotype. While such strategies may result in improved traits for simple phenotypes, they do not address how numerous mutations or genetic changes in combination may be required for improvements in complex phenotypes. As organic acid inhibition affects many different cellular processes, acid tolerance is one such phenotype that can be conferred to varying degrees by many different combinations of genetic changes.

Organic acids have been historically utilized as preservatives in both food and feed products due to their inhibitory properties to various microorganisms (5). Although the antimicrobial properties of these organic acids have been widely exploited commercially, the observed growth inhibition also stands as a critical hurdle in the development of economical bioprocesses. Organic acid-related growth inhibition has several reported modes of action, including disturbance of cytoplasmic pH, anion accumulation within the cytosol at high organic acid titers, and increasing osmotic stress due to addition of neutralizing agents along the fermentative time course (6,7). While pH deviations are minimized in controlled fermentations, anion accumulation and osmotic stresses continue to challenge metabolic engineering efforts in the development of highly productive host strains (8). The rest of this chapter will focus on alleviating the anionic effects of organic acid stress toward the development of a robust production strain.

Anion accumulation inside the cell can cause growth defects in a number of ways. Explicitly, undissociated weak acids are able to diffuse freely through the cell membrane where they release a proton and subsequently lower cytosolic pH (9). Significant accumulation of anions in the cytosol results in increased osmotic stress and a corresponding increase in free potassium (10). The physiological response to balance anionic concentrations and maintain a constant turgor pressure is to increase export of glutamate, resulting in reduced growth and viability of the cell (11). Additionally, the anion itself can result in increased inhibition of metabolic pathways characterized by severe growth defects. For example, measurements of the intracellular metabolite pools under acetate stress indicate a significant increase in homocysteine pools accompanied by a reduction in the downstream methionine pools, implying inhibition at this step in methionine biosynthesis. Further, methionine supplementation has been shown to relieve growth inhibition due to acetate levels as high as 8 mM (11). Such inhibitory effects are anion specific and often hard to elucidate across the complex metabolic networks. These findings from Roe et al. (11) surrounding the metabolic inhibition specific to acetate stress helped to form the hypothesis that similar metabolic inhibition may be the basis for the apparent growth inhibition specific to 3-HP.

7.2   UNDERSTANDING 3-HP TOXICITY

The work described herein focuses on improving the understanding of organic acid tolerance mechanisms, specifically for 3-HP, in *Escherichia coli*. Biosynthetic processes yielding 3-HP have previously been demonstrated from development of recombinant hosts (12,13). However, as mentioned previously, severe growth inhibition has been observed for extracellular acid levels as low as 20 g/L in minimal media (pH 7.0), which severely impacts the economic feasibility of 3-HP production as a platform chemical. Furthermore, engineering a tolerant host for production has proven complex with numerous genetic targets and integrated metabolic networks incorporated into an overall toxicity profile.

7.2.1   Choosing an Approach for Evolving Tolerance

Traditional approaches to strain engineering typically employ either informed metabolic engineering methods (see Chapters 2 and 3), which are reliant on previous understanding of genetic function, or directed evolutionary approaches (see Chapter 4), which are based on application of a selective pressure on a genetically diverse population to identify previously unknown or uncharacterized genotypes (Figure 7.1). Although directed evolution approaches do provide insight into the unknown, they are oftentimes laborious and qualitative, and result in an incomplete understanding of the engineered phenotype. Further, since phenotypes such as growth or productivity are a function of numerous unknown factors, successful strain selections have primarily relied on an iterative mutation strategy, which simultaneously result in the accumulation of deleterious mutations. Recent reports have disclosed the use of extrachromosomal or disruptional mutagenesis approaches that are meant to address the concerns with traditional directed evolution (14,15). However, these approaches still require substantial follow up to identify exact causal linkages between a given phenotype and the underlying genotype.

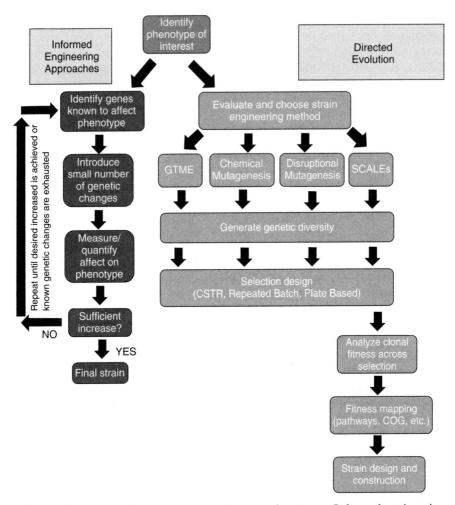

FIGURE 7.1. Sample flowchart for evolving new phenotypes. Informed engineering approaches are displayed in dark gray shaded boxes, whereas directed evolutionary approaches are illustrated in light gray shaded boxes. GTME, global transcription machinery engineering; CSTR, continuous stirred tank reactor; COG, clusters of orthologous groups.

For this study we chose to employ multi-SCale Analysis of Library Enrichment (SCALEs). This approach involved simultaneous growth selections on mixtures of multiple plasmid libraries containing defined, yet unique, insert sizes (or scales). Through microarray and multiscale analysis, the signal contribution and associated fitness (W) of each of the distinct, different sized libraries was identified. Fitness, in this case, is defined as the enrichment of each region in the selected population over time ($W = \ln(X_{i,t}/X_{i,t-1})$). This resulted in the accurate identification of the location and size of the fitness-altering loci that contributed to the desired phenotype. For a detailed description of the procedures involved in the multiscale analysis of library enrichment, the reader is referred to Lynch et al. (16).

The model system described herein was focused on the identification of clones exhibiting increased growth rate in the presence of 3-HP. Our studies first involved the use of the SCALEs method to characterize the genotypes and phenotypes under selection in a continuous culture selection. The information from this analysis was then used to design a new selection directed more precisely at 3-HP tolerance phenotypes. Finally, a receiver operator curve (ROC) analysis, enabled by SCALEs, was employed to measure and compare the sensitivity and selectivity of both selections and provide a unique insight into the complexities of selection design (16–19).

7.2.2 Selection Design for Evolving 3-HP Tolerance

Continuous culture methodologies, such as the use of chemostats and turbidistats, are traditionally employed for strain selection in the presence of differing levels of a toxic compound (18–21). We chose to first employ this approach as a base case to select for clones exhibiting increased 3-HP tolerance from our *E. coli* plasmid-based genomic library (18). The starting population for selection was comprised of five representative *E. coli* K12 genomic libraries that were transformed into MACH1-T1^R (18) and recovered until mid-log phase (OD_{600} ~0.2) under microaerobic conditions. Greater than 10^6 library clones were then introduced to the continuous flow reactor (CFR) and cultured for 60 hours, at which point the selection was stopped due to significant biofilm growth on the reactor walls. The CFR was fed with 3-(N-morpholino)propanesulfonic acid (MOPS) minimal media blended with increasing levels of 3-HP (pH = 7.0, 0–20 g/L). The CFR was maintained similar to a common turbidistat, in that the dilution rate was constantly adjusted in an effort to maintain an approximate cell density of 10^7 cells/mL and to avoid a nutrient-limited environment. Samples of the population were acquired at approximately 30, 40, and 50 doublings (corresponding to 100-, 1000-, and 10,000-fold enrichment, respectively) and were analyzed via microarray and decomposed according to the SCALEs methodology (16). By performing this analysis we were able to quantify the concentration of each library clone (X_i), or member of the library, maintained within the plasmid library throughout the selection. Following this quantification, we were able to map the genome-wide fitness and enrichment patterns for the continuous flow selections in the presence of 3-HP at each time point.

To further characterize the efficacy of the selection for identification of 3-HP tolerant clones, we chose 17 clones to observe for further testing. Clones were obtained either via sequencing individual clones isolated from selection samples or conventional molecular cloning of regions corresponding to significantly increased fitness from the SCALEs analysis. These clones were then introduced individually into batch cultures with minimal media and 20 g/L of neutralized 3-HP. Growth was monitored as a function of OD_{600} over a 24-hour period and specific growth rates were calculated for regions of growth corresponding to minimal doubling times. For this particular selection, only 53% of clones tested (9 of the 17 tested) demonstrated significantly increased fitness in the presence of inhibitory levels of 3-HP. As expected, these results confirm that the CFR selection not only resulted in enrichment of 3-HP tolerant clones, but was also selected for other various phenotypes that may have increased residence time with the reactor, such as wall adherence and/or biofilm formation.

In an effort to better quantify these observations, we performed an ROC analysis (17). ROC curves have traditionally been used in signal detection theory to compare the predictive power of a model as the criterion, or thresholds, are varied. ROC curves plot the true positive rate (sensitivity) against the false positive rate (1-specificity), which allows for the rapid assessment of the predictive power of a test and, importantly, to compare the utility of multiple tests.

$$\text{Sensitivity} = \text{True Positive Rate (TPR)}$$
$$= \text{true positives}/(\text{true positives} + \text{false negatives})$$

$$\text{Specificity} = \text{True Negative Rate (TNR)}$$
$$= \text{true negatives}/(\text{false positives} + \text{true negatives})$$

We used this technique to assess how efficiently our CFR selections identified clones with 3-HP tolerant phenotypes. True positives were quantified from clones that were identified by the SCALEs method with significant fitness gains that were separately confirmed to grow faster than the control at elevated 3-HP concentrations. Similarly, false positives were quantified from clones characterized by increased fitness as identified by SCALEs that did not relate to an observed increased growth in the presence of 3-HP. Finally, the number of true negatives was set according to clones with fitness values below a given cutoff that did not show increased tolerance, while false negatives were those with low fitness values and an increased growth rate (Figure 7.2).

The advantage to a genome-wide analysis, such as SCALEs, in this type of evaluation is the ability to gain information about the true and false negatives, which are essential for a quantitative ROC analysis. These data are unique when compared with a traditional library selection followed by a sequencing-based output in which nothing can be determined about the unsequenced population. A measure of the overall quality of a test can be provided by assessing the area under the ROC curve (AUC), where larger values correspond to tests that increase in true positives relative to any increase in false negatives.

In all cases the CFR selections produced ROC curves above the $x = y$ line, indicating that gains in true positives were not accompanied by an equivalent increase in false negatives (see Warnecke et al. (18) for detailed ROC diagrams). This result confirms that 3-HP tolerance was under selection in our CFR. However, the AUC values for the CFR selection maintain a constant

	HIGH Fitness According to SCALEs	LOW Fitness According to SCALEs
INCREASED Tolerance (Measured in the Lab)	*TRUE POSITIVE*	*FALSE NEGATIVE*
NO Tolerance (Measured in the Lab)	*FALSE POSITIVE*	*TRUE NEGATIVE*

FIGURE 7.2. Quantifying true and false negatives and positives in an ROC analysis. Following the SCALEs analysis, genes were categorized by fitness (W) as having positive or negative effect on 3-HP tolerance and confirmed as a true positive or negative by experimentation to confirm increased growth rate in the presence of 3-HP.

value of approximately 0.77, suggesting that selective pressure specific to 3-HP tolerance is stagnant after the initial sample. This finding is counterintuitive in that selective pressure should have increased as increasing amounts of 3-HP were introduced to the growth media. In actuality, the frequency of clones contributing to the formation of biofilm communities resulted in an increased false positive rate, thus keeping the AUC constant. This finding implies that selective pressure was being increasingly driven by wall adherence as opposed to 3-HP tolerance.

7.2.3 Taking a Closer Look at Selection Design

Based on this analysis, we began a second selection where improvement would be measured by an increase in the AUC resulting from a comparable ROC analysis. This selection was designed in order to enhance the specificity of the selection toward increased specific growth rate in the presence of 3-HP and not for increased biofilm formation or other potentially selectable phenotypes such as decreased lag time and increased final density. We chose to utilize serial batch selections with a progressively decreasing concentration of 3-HP in an effort to both reduce clones with improved fitness resulting from wall adherence and to increase the corresponding true positive rate throughout the selection in contrast to the stagnant sensitivity observed for the CFR. The design of selection varied from the original based on the inhibitory, rather than bactericidal, effect that 3-HP has on growth rate. Specifically it can be assumed that high levels of 3-HP will inhibit growth of all but the most tolerant clones, or the true positives, at the onset of selection. This initial growth inhibition provides the selective pressure required for the fastest-growing clones to become enriched relative to clones that are more sensitive to 3-HP. Under the reverse scenario of an increasing gradient of 3-HP, all clones capable of growing at an initially low concentration of 3-HP will have an opportunity to become enriched over the entire period of selection. Additionally, those clones exhibiting mutations by random genetic drift will also be enriched greatly toward the culmination of the selection. Some of such clones may grow rapidly at low concentrations but not at all at high concentrations. These clones would be considered false positives. In terms of the ROC analysis, selection using a decreasing 3-HP concentrations gradient should result in an increase in specificity (a decrease in false negative rate) accompanied by an increase in sensitivity (true positives).

A second selection, employing successive batch cultures with decreasing levels of 3-HP (20 to 0 g/L), was designed. Following each batch, transfer samples were taken and the plasmid population was isolated, labeled, and hybridized to DNA microarrays. Microarray data were further examined according to the SCALEs methodology. To test for increased sensitivity and specificity, we chose 20 unique clones from the serial dilution (SD) selections by isolating and sequencing of individual clones from enriched pools or by molecular cloning of the highly enriched clones identified by the SCALEs data. We used this clonal subset to generate data to perform an ROC analysis used to quantitatively compare the two selections. As was expected, 0% of the clones obtained from the SD expressed biofilm phenotypes, whereas 88% of the clones obtained from the 60-hour CFR sample showed this phenotype. The clones isolated from the SD selection were then used to evaluate growth rates in the presence of 20 g/L 3-HP. We found that 100% of these clones had a statistically improved growth rate compared with only 53% of the clones evaluated from the CFR selection.

These data were then used to generate a ROC curve, which showed a quantifiable improvement in the SD selection when compared with the original CFR data (the reader is again referred to Warnecke et al. (18) for the detailed ROC diagram). More specifically, the AUC increased with each successive sample for the SD selection, indicating that selection for 3-HP tolerance phenotypes was maintained throughout the experiment. This is in contrast to the CFR selection, which maintained a more stagnant AUC profile throughout. Further, the SD selection shows a 17% overall increase in AUC when compared with the cumulative CFR selection, implying that the SD selection more effectively identified true 3-HP tolerant clones compared with the CFR selection.

7.2.4   Constructing the 3-HP Toleragenic Complex

As shown above, we have performed a selection that showed a strong correlation between fitness and the desired 3-HP tolerant phenotype. The clone-specific fitness was calculated according to the SCALEs analysis for the optimized selection described above. This fitness was then further segmented to assign fitness values to the individual genes contained within the clone's insert DNA. Genes and their corresponding fitness measures were then grouped according to their associated metabolic pathways. This analysis allowed for the assessment of the fitness of, or enrichment for, specific metabolic pathways. The distribution of pathway enrichment values revealed a clear pattern of differentiation across all metabolic pathways considered.

Segmenting the fitness according to metabolic pathway demonstrated that 3-HP related fitness was conferred by increased dosage of genes from only a handful of metabolic pathways. The identified metabolic pathways that accounted for the most significant improvement in overall fitness included chorismate, threonine–homocysteine, arginine–polyamine, and nucleotide biosynthesis superpathways. These pathways and their interactions mostly comprise what we have termed the 3-hydroxypropionic acid toleragenic complex (3-HP-TGC). The metabolic complex, shown in Figure 7.3, comprises all of the genes identified by the SCALEs approach as contributing to metabolic processes and increased tolerance to 3-HP (22).

This complex has been confirmed both by supplementation of media metabolites from the complex and by genetic modifications of the 3-HP-TGC (Figure 7.4). These data illustrate the importance of the selection strategy on the identification of a phenotype of interest, in this case increased growth rates in the presence of the growth inhibitor 3-HP. More specifically, it would not be possible to construct meaningful networks for a selection correlated to a

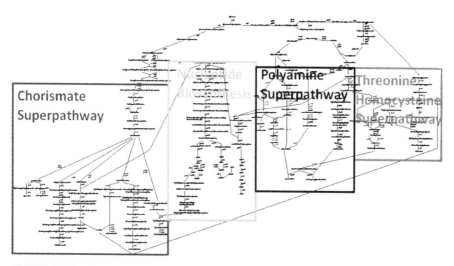

FIGURE 7.3. The 3-HP toleragenic complex (3-HP-TGC) as constructed from metabolic pathway fitness data. Subsections of the 3-HP-TGC are denoted for the chorismate, nucleotide biosynthesis pathway, polyamine, and threonine/homocysteine superpathways. *(See insert for color representation of the figure.)*

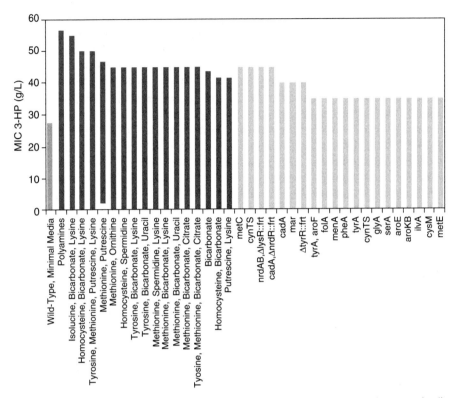

FIGURE 7.4. Confirmation of 3-HP tolerance corresponding to supplements (red) and genetic modifications (green). Tolerance was quantified as the minimum inhibitory concentration (MIC) of 3-HP in triplicate ($n = 3$) at pH = 7.0. *(See insert for color representation of the figure.)*

high false positive rate. In particular, decreasing the selective pressure over time ensures that enrichment occurs for clones with even a small selective advantage.

7.3   STRAIN DESIGN

7.3.1   Evaluation of the 3-HP-TGC

To better evaluate the toleragenic complex, we divided the 3-HP-TGC into component pathways, or gene groupings, surrounding highly connected nodes involved in the homocysteine, chorismate, polyamine, lysine, uracil, and citrate synthesis pathways. Each of the component pathways was then used to high-light genetic elements at key toxicity points as well as potential supplemental strategies to overcome limitations. Toxicity evaluations were carried out involving the identified supplements, genetic modifications, and combinations designed to evaluate the groupings of multiple branches of the 3-HP-TGC.

The supplement data gathered according to the combinatorial strategy above showed the significant impact that saturation of the polyamine pathway, by addition of the entire polyamine group (composed of putrescine, spermidine, ornithine, citrulline, bicarbonate, and glutamine), has on the overall tolerance to 3-HP. As such, it was important to design a strain with an emphasis on optimization of polyamine production to alleviate a large fraction of growth inhibition. Additionally, increased tolerance (as demonstrated by a greater than 200% increase in minimum inhibitory concentration [MIC]) was noted for supplementation with one supplement from each of the chorismate, homocysteine, polyamine, and lysine pathways suggesting that modifications from each of the pathways should be included in the final tolerant strain design.

The results presented above illustrate the foundation of technology that has been developed in our laboratories that can be applied to explore complex phenotypes such as solvent tolerance. The ability to fully characterize selection dynamics toward the identification of a single desired phenotype is critical for the project proposed herein. Further development of this high-resolution, genome-wide technology platform is ongoing and will enable the rapid evaluation of desirable phenotypes.

7.3.2 Complex Tolerant Phenotype: Metabolism of 3-HP to a Toxic Intermediate

Evolving improved tolerance of the 3-HP production host is complex in that it includes developing increased resistance to not only the final 3-HP product, but also to potential toxic intermediates or degradation products. In the case of the OPXBIO 3-HP production route, preliminary fermentations with 3-HP production strains demonstrated significant metabolic conversion of 3-HP to 3-hydroxypropionaldehyde (3-HPA), which resulted in both decreased 3-HP yield and increased toxicity toward the production strain. 3-HPA, also known as reuterin, is an antibiotic naturally produced by Lactobacillus reuteri and is commonly used as a biopreservative in food products (23). 3-HPA exists in a three-way equilibrium between the hydrate, the dimer, and acrolein. Although the exact antimicrobial mechanism of reuterin has not been elucidated to date, the proposed targets include the sulfhydryl enzymes (24).

A review of the SCALEs results described above identified a particular genetic network composed of 22 aldehyde dehydrogenases (ALDs), which demonstrated a low level of enrichment throughout the serial-dilution selection $(1 < W < 2)$. To further investigate this apparent metabolism, assays were developed to quantify 3-HPA/3-HP for various ALD deletion strains cultured with exogenous 3-HP. One particular ALD identified in the SCALEs data set, *puuC*, had been previously reported to catalyze the conversion of 3-HP to 3-HPA (25). Further experimentation with *ΔpuuC* base strains showed a significant decrease in 3-HP conversion to the toxic aldehyde, which thereby improved overall tolerance to the host.

The *puuC* findings have further implications when applied to the understanding of improved tolerance to 3-HP. Specifally, *puuC* has been characterized as a γ-glutamyl-γ-aminobutyraldehyde dehydrogenase involved in polyamine degradation that is induced by increased levels of polyamines. The observed conversion to 3-HPA in the presence of 3-HP implies that increased polyamine pools may be inducing this conversion mechanism. This mechanism was further established by quantification of increased 3-HPA/3-HP in the presence of supplemental polyamines such as putriscine. The polyamine pathway is critical for increased tolerance to 3-HP due to the global effects on the overall metabolic network as illustrated in the toleragenic complex above. Polyamines are synthesized in the cell as natural osmolarity stabilizers and are required for numerous cellular processes. While *E. coli* do produce polyamines naturally, the basal levels are not sufficient for normal cellular activities under conditions of 3-HP stress. More specifically, 3-HP can act as a potent chelator, particularly at high concentrations, causing the cell to sense a hypo-osmolar state, which can be countered by overproduction of various polyamines. Interestingly, overexpression of polyamine biosynthetic genes alone is not able to counter the toxic effects of 3-HP without balancing accompanying side effects such as conversion to the toxic aldehyde. Additionally, increased production of polyamines can lead to the accumulation of carbamoyl-phosphate, a precursor to polyamine synthesis, which degrades to the toxic byproduct, cyanate. As such, it is important to balance the increased polyamine levels with increased expression of a cyanase, which is capable of minimizing accumulation of cyanate within the cell.

Figure 7.5 illustrates the MIC of 3-HP on wild-type *E. coli* with and without supplementation of various polyamines (putrescine, cadaverine, and spermidine), as well as an isolate overexpressing a cyanase. By applying these findings in concert with the other components of the 3-HP-TGC, it has been possible to engineer *E. coli* to tolerate 3-HP levels as high as 100 g/L.

7.4 COMBINING 3-HP TOLERANCE AND 3-HP PRODUCTION

Engineering strains for industrial production requires stacking multiple complex phenotypes (4). Three production strains were constructed with various genetic modifications that correspond to increased 3-HP tolerance corresponding to the 3-HP-TGC for further evaluation. Specifically, we were interested in evaluating the impact that each of the tolerance modules had on production metrics such as final titer and specific productivity. The results for the three tolerance modules are displayed in Figure 7.6. As shown below, each of the three tolerant strains resulted in significant improvements in both 3-HP

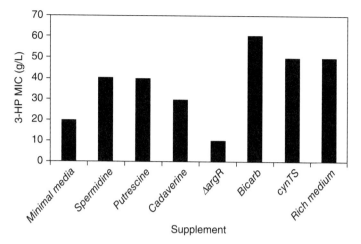

FIGURE 7.5. MICs for 3-HP with and without supplementation with the polyamines (putrescine, cadaverine, and spermidine) as well as overexpressing a cyanase. MICs were performed in triplicate ($n = 3$) at pH = 7.0.

titer and specific productivity, and in the case of tolerant strain 3, the specific productivity was increased by 10-fold compared with the control.

7.5 SUMMARY

The work described in this chapter was focused on improving the understanding of organic acid tolerance mechanisms, specifically for increased production of 3-hydroxypropionic acid (3-HP), in *E. coli*. Successfully evolving complex phenotypes, such as organic acid tolerance, relies on approaches capable of generating a global understanding of toxicity in an effort to utilize synergistic effects. In this chapter, we described how to design a selection that shows a strong correlation between fitness and the desired 3-HP-tolerant phenotype and how the application of a genome-wide approach such as SCALEs allowed for the rapid identification of numerous genetic changes. The results of our studies identified hundreds of genes and other genetic elements that when at increased copy confer varying levels of tolerance to the presence of 3-HP in *E. coli*. When applied alone, these genetic changes may allow for small increases in tolerance; but when applied together they allow for insight into the 3-HP toxicity mechanisms. By grouping genetic elements that confer tolerance by their metabolic roles, we were able to identify key metabolic pathways that are inhibited by 3-HP and to increase productivity by overcoming 3-HP inhibition.

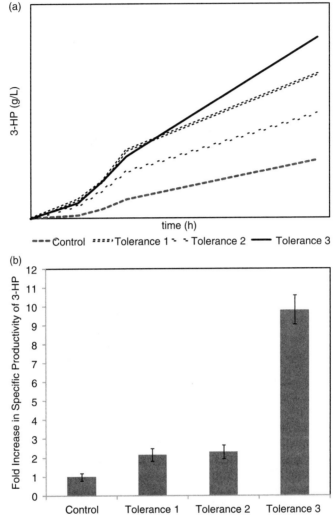

FIGURE 7.6. The effects of improved tolerance on 3-HP production: (a) improved titers and (b) improved rates with tolerance modifications incorporated.

REFERENCES

1. Werpy, T., Petersen, G., Aden, A., Bozell, J., Holladay, J., White, J., Manheim, A., Elliot, D., Lasure, L., Jones, S., Gerber, M., Ibsen, K., Lumberg, L., and Kelley, S. (2004) Top value added chemicals from biomass. Oak Ridge, TN, U.S. Department of Energy, Washington, DC. Volume 1: Results of Screening for Potential Candidates from Sugars and Synthetic Gas.

2. Berg, I.A., Kockelkorn, D., Buckel, W., and Fuchs, G. (2007) A 3-hydroxypropionate/4-hydroxybutyrate autotrophic carbon dioxide assimilation pathway in archaea. *Science*, **318**, 1782–1786.

3. Herter, S., Farfsing, J., Gad'On, N., Rieder, C., Eisenreich, W., Bacher, A., and Fuchs, G. (2001) Autotrophic CO2 fixation by Chloroflexus aurantiacus: study of glyoxylate formation and assimilation via the 3-hydroxypropionate cycle. *J Bacteriol*, **183**, 4305–4316.

4. Patnaik, R. (2008) Engineering complex phenotypes in industrial strains. *Biotechnol Prog*, **24**, 38–47.

5. Skrivanova, E. and Marounek, M. (2007) Influence of pH on antimicrobial activity of organic acids against rabbit enteropathogenic strain of Escherichia coli. *Folia Microbiol (Praha)*, **52**, 70–72.

6. Booth, I.R. (1999) The regulation of intracellular pH in bacteria. *Novartis Found Symp*, **221**, 19–28.

7. Russell, J.B. and Diez-Gonzalez, F. (1998) The effects of fermentation acids on bacterial growth. *Adv Microb Physiol*, **39**, 205–234.

8. Andersson, C., Helmerius, J., Hodge, D., Berglund, K.A., and Rova, U. (2009) Inhibition of succinic acid production in metabolically engineered Escherichia coli by neutralizing agent, organic acids, and osmolarity. *Biotechnol Prog*, **25**, 116–123.

9. Warnecke, T. and Gill, R.T. (2005) Organic acid toxicity, tolerance, and production in Escherichia coli biorefining applications. *Microb Cell Fact*, **4**, 25.

10. McLaggan, D., Naprstek, J., Buurman, E.T., and Epstein, W. (1994) Interdependence of K+ and glutamate accumulation during osmotic adaptation of Escherichia coli. *J Biol Chem*, **269**, 1911–1917.

11. Roe, A.J., O'Byrne, C., McLaggan, D., and Booth, I.R. (2002) Inhibition of Escherichia coli growth by acetic acid: a problem with methionine biosynthesis and homocysteine toxicity. *Microbiology*, **148**, 2215–2222.

12. Suthers, P.F. and Cameron, D.C. (2001) Production of 3-Hydroxypropionic acid in recombinant organisms. Wisconsin Alumni Research Foundation. PCT WO 01-16346(6852517).

13. Vollenweider, S. and Lacroix, C. (2004) 3-hydroxypropionaldehyde: applications and perspectives of biotechnological production. *Appl Microbiol Biotechnol*, **64**, 16–27.

14. Alper, H., Moxley, J., Nevoigt, E., Fink, G.R., and Stephanopoulos, G. (2006) Engineering yeast transcription machinery for improved ethanol tolerance and production. *Science*, **314**, 1565–1568.

15. Gill, R.T., Wildt, S., Yang, Y.T., Ziesman, S., and Stephanopoulos, G. (2002) Genome-wide screening for trait conferring genes using DNA microarrays. *Proc Natl Acad Sci U S A*, **99**, 7033–7038.

16. Lynch, M.D., Warnecke, T., and Gill, R.T. (2007) SCALEs: multiscale analysis of library enrichment. *Nat Methods*, **4**, 87–93.

17. Fawcett, T. (2006) An introduction to ROC analysis. *Pattern Recognit Lett*, **27**, 861–874.

18. Warnecke, T.E., Lynch, M.D., Karimpour-Fard, A., Sandoval, N., and Gill, R.T. (2008) A genomics approach to improve the analysis and design of strain selections. *Metab Eng*, **10**, 154–165.

19. Warnecke, T.E., Lynch, M.D., Karimpour-Fard, A., Lipscomb, M.L., Handke, P., Mills, T., Ramey, C.J., Hoang, T., and Gill, R.T. (2010) Rapid dissection of a complex phenotype through genomic-scale mapping of fitness altering genes. *Metab Eng*, **12**(3), 241–250.

20. Patnaik, R., Louie, S., Gavrilovic, V., Perry, K., Stemmer, W.P., Ryan, C.M., and del Cardayré, S. (2002) Genome shuffling of Lactobacillus for improved acid tolerance. *Nat Biotechnol*, **20**, 707–712.

21. Dykhuizen, D.E. and Hartl, D.L. (1983) Selection in chemostats. *Microbiol Rev*, **47**, 150–168.

22. Gill, R.T., Lipscomb, T.E., and Lynch, M.D. (8-19-2010) Compositions and methods for enhancing tolerance for the production of organic acids produced by microorganisms. (US2010/0210017 A1).

23. Rasch, M. (2002) The influence of temperature, salt and pH on the inhibitory effect of reuterin on Escherichia coli. *Int J Food Microbiol*, **72**, 225–231.

24. Cleusix, V., Lacroix, C., Vollenweider, S., Duboux, M., and Le Blay, G. (2007) Inhibitory activity spectrum of reuterin produced by Lactobacillus reuteri against intestinal bacteria. *BMC Microbiol*, **7**, 101.

25. Jo, J.E., Raj, S.M., Rathnasingh, C., Selvakumar, E., Jung, W.C., and Park, S. (2008) Cloning, expression, and characterization of an aldehyde dehydrogenase from Escherichia coli K-12 that utilizes 3-hydroxypropionaldehyde as a substrate. *Appl Microbiol Biotechnol*, **81**, 51–60.

8

COMPLEX SYSTEM ENGINEERING: A CASE STUDY FOR AN UNSEQUENCED MICROALGA

MICHAEL T. GUARNIERI, LIEVE M.L. LAURENS, ERIC P. KNOSHAUG, YAT-CHEN CHOU, BRYON S. DONOHOE, AND PHILIP T. PIENKOS

8.0 HISTORICAL PERSPECTIVE

In 1978, the U.S. Department of Energy (DOE) initiated the Aquatic Species Program (ASP), which was managed by the National Renewable Energy Laboratory (NREL, known at the time as the Solar Energy Research Institute). The purpose of the ASP was to evaluate the potential of non-terrestrial crops to serve as feedstocks for biofuel production. Initially the scope of the investigation included microalgae, cyanobacteria, macroalgae, and wetland emergents, and the potential products included hydrogen, lipids, ethanol and other alcohols, syngas, and pyrolysis fluids. The intermediates could be used in fuel cells or upgraded to a variety of fuels including biodiesel, renewable diesel, renewable gasoline, and renewable jet fuel. Very quickly the target organisms were downselected to eukaryotic microalgae primarily because of anticipated productivity of algae, the high energy density of lipids, and the ease of conversion to biodiesel.

For 18 years, from 1978 to 1996, DOE funded the ASP (through significant swings in annual budgets) with most of the work carried out by academic subcontractors with NREL providing the overall project management. During that period, all aspects of the value proposition were investigated, ranging from basic algal biology through large-scale cultivation, harvest, dewatering, conversion to fuels, and techno-economic analysis. Many breakthroughs

Engineering Complex Phenotypes in Industrial Strains, First Edition. Edited by Ranjan Patnaik.
© 2013 John Wiley & Sons, Inc. Published 2013 by John Wiley & Sons, Inc.

were achieved including the assembly of a 3000-strain culture collection, development of a capital- and energy-minimized raceway pond for cultivation, continuous operation of two $1000\,m^2$ raceways for a year in Roswell, New Mexico, and production of biodiesel samples for testing. The best results and assumptions were used to estimate capital and operating costs. Optimistic assumptions led to a conclusion that algal biofuels could be produced at a cost of \$40–60 per barrel. Crude oil prices at the time held at around \$20 per barrel and most projections suggested that the price would remain at that level for decades to come. Facing budget challenges and recognizing the long-term nature of R&D needed to bring algal biofuels to production costs that could compete with petroleum, DOE decided to terminate the ASP in 1996 to focus on cellulosic ethanol. A comprehensive closeout report on the microalgae work done for the ASP was released in 1998 (http://www.nrel.gov/biomass/pdfs/24190.pdf). The detail and comprehensive nature of this report have provided great value in the intervening years for researchers seeking to enter this field.

In 2005, DOE and the U.S. Department of Agriculture (USDA) released a report, commonly known as the "Billion Ton Study" (http://feedstockreview.ornl.gov/pdf/billion_ton_vision.pdf), which provided for the first time estimates of the amount of terrestrial biomass that could be sustainably harvested on a yearly basis. This upper limit provided guidance for our national capacity for biofuels, somewhere in the range of 40–60 billion gallons gasoline equivalents (gge), depending on the yields of biofuel that could be derived from biomass. Considering that the United States burns approximately 140 billion gallons of gasoline, 40 billion gallons of diesel fuel, and 25 billion gallons of jet fuel annually, this limited capacity for domestic biofuel production was clearly inadequate to provide energy security. Also considering that the only biofuel that was being considered from terrestrial biomass (excluding food crops such as corn or soybeans) in 2005 was ethanol, it was clear that biofuel replacements for high energy density fuels such as diesel and jet fuel would require a different feedstock.

With that in mind, researchers at NREL began to reconsider the assumptions that led to the close of the ASP. Among the changes that had occurred in that 10-year period were increases and increased volatility in the price of crude oil, increased awareness of the role of CO_2 emissions in global climate change, increased demand for energy security, and increased sensitivity to the food versus fuel debate. At this point it became apparent that algal biofuels could be a game changer because algae could capture and recycle CO_2 directly from fixed sources such as power plants, steam methane reformers, cement kilns, and other large-scale contributors to CO_2 emissions. Algae could also be cultivated on nonproductive lands that would not even support growth of energy crops such as switchgrass and miscanthus. And algae could be cultivated with brackish or saline water taken from marine sources or from saline aquifers or oil well-produced water. And so, in 2006, NREL began a strategic initiative to look for ways to revive its algal biofuels R&D effort.

In addition to industrial outreach, NREL built a strategy around small projects funded either internally through the national lab Laboratory Directed Research and Development (LDRD) program, which makes a percentage of overhead funds available for competitive research proposals. Other funding opportunities also became available, including one from the Air Force Office of Scientific Research, initiating a program to fund basic research in algal biology with a focus on lipid production. None of these funding mechanisms could match the peak funding days of the ASP, but taken together, they provided a vehicle for NREL to re-establish itself as a leader in algal biofuels R&D.

It must be noted that this began to take place at a time of phenomenal growth in interest in algal biofuels, a time in which many new companies were being formed to commercialize algal biofuels. It was essential that the limited funding available from these sources be put to work in the most effective manner. After evaluating the overall landscape we determined that our biggest contributions could be made in algal biology and algal compositional analysis, and so our proposals and subsequent projects focused on these two areas. To further focus our efforts and to leverage the funding from different sources, we chose a single organism, *Chlorella vulgaris*, to be our model organism, to be used in as many projects as we could. We chose *C. vulgaris* for a number of reasons:

- Rapid growth and high lipid content.
- Large-scale cultivation history for production of nutritional supplements.
- Reports of successful genetic transformation.

There were also a number of drawbacks for this choice:

- Small cell size and very tough, uncharacterized cell wall made lipid extraction and cell lysis difficult.
- The genome sequence was not available, although related strains were in the queue at the Joint Genome Institute.
- *C. vulgaris* is a fresh water strain, which would raise issues for sustainable large-scale production for biofuels.

Ultimately, we determined that the positive attributes outweighed the negative and began working with *C. vulgaris* as our model organism. The challenges, though real, would provide guidance for research groups who also wished to work with unsequenced strains. The freshwater adaptation of *C. vulgaris* was certainly a major issue, but other groups were also developing freshwater strains for production. In addition, it had been shown that it is possible to adapt fresh water strains for growth in brackish or saline water, and so we decided to focus our work on *C. vulgaris*.

As a result, in 2008 we assembled a project portfolio using *C. vulgaris*, exploiting the synergy that would develop from connecting a number of small projects and otherwise individual researchers with a common single model organism. It was clear from the outset that strain improvement based on increased lipid production would be challenging. The lipid metabolic pathways are well known and have been investigated in plants for improved productivity in oil seed crops (1) and in animals for drug discovery (2). But the pathways for storage lipid production and metabolism are complicated by the common steps involved in production of storage lipids (triacylglycerides [TAGs]) and membrane lipids (phospholipids, sulfolipids, and glycolipids). Some successes have been reported for altered flux into TAG synthesis in algae through both classical selection for mutants resistant (see Chapter 1) to herbicides, which block lipid biosynthetic steps (3–4), and, more recently, fluorescence-activated cell sorting (FACS) enrichments using the lipid-specific dye Nile Red (5). Strain improvements based on genetic engineering of algae had been attempted during the aquatic species program (6), which saw successful overexpression of the ACCase gene but no increases in lipid production. In the past two years, several papers have been published describing alteration of specific gene expression levels with mixed success at improving lipid production in *Chlamydomonas reinhardtii* (7–8) and unpublished reports indicate that further improvements are in hand, but it is widely recognized that *C. reinhardtii* is not sufficiently robust to serve as a production strain. And so we have been encouraged to pursue this path in a strain that begins with better production characteristics.

8.1   ANALYSIS OF ALGAL BIOMASS COMPOSITION

8.1.1   Defining the Parameters of an "Ideal" Strain

A techno-economic analysis of the biofuels production process made it clear that overall lipid productivity of the algal production strain is intricately linked with the economics of the production process (9). A sensitivity analysis of the techno-economic model of microalgae for fuel production showed that doubling and halving the lipid content of the algae used caused a $4 decrease and an $8 increase, respectively, around the benchmark of $9.64 per gallon of biofuel. This analysis reveals lipid content as the most important determinant of the algal biofuels process economics, followed closely by biomass production and growth rate of the algae (9). Furthermore, the lipid content and composition of algae has been one of the challenges, but also drivers for process development and accelerated algal biofuels research in the past couple of years (10). In light of these economic implications, it is necessary to have an accurate, objective measure of lipids in algae. In addition to having an accurate lipid content determination, there is also an emphasis on obtaining this measure rapidly, in a high-throughput manner to enable screening of a large number of strains (see Chapter 5). Besides lipids, algal biomass also contains proteins and carbohydrates, and the ratio and composition of these compounds will further determine the fate of the residual algal biomass after the lipids have been extracted, which can play a significant role in the overall process, perhaps even drive the development of alternative uses for the residual algal biomass; for example, high fermentable sugar content of the residue could be converted into fuel ethanol.

8.1.2 Tool Development for the Analysis of Growth and Lipid Production

A robust tool for measuring lipid in algae has to provide an accurate, precise, and reproducible measure of total lipids and preferably can be carried out on a small scale (in the mg of biomass range, requiring milliliter volumes of culture rather than liter volumes). The tool also has to be species agnostic, in that the measurement cannot be dependent on the algae type subjected to the analysis; for example, susceptibility of the method to cell wall permeability is unacceptable. A survey of the literature indicates a wide range in reported lipid contents, which can be traced back to a vague definition of lipids and the inherent variability of the lipid content in algae throughout the growth phase of a culture.

It is well known that physiological conditions will influence the composition and the total lipid content of algae. Lipid content and fatty acid (FA) composition vary considerably during the growth cycle. Algal lipids from cultures that are in the exponential growth phase consist mainly of polar lipids, such as phospho- and glycolipids, which make up the cell organelle and photosynthetic membranes. In many algal species, an increase in TAGs is observed during the stationary phase. For example, in the chlorophyte *Parietochloris incise*, TAGs increased from 43% (of the total FAs) in the logarithmic phase to 77% in the stationary phase (11), and in the marine dinoflagellate *Gymnodinium* sp., the relative amount of TAGs increased from 8% during the logarithmic growth phase to 30% during the stationary phase.

In the context of quantification of lipids, this variation in lipid composition over the culture's growth period leads back to the problem with the definition of lipids as molecules more soluble in organic solvents than in water (12). In agreement with this definition, traditional lipid quantification is based on the gravimetric solvent extraction yield. However, the wide variety of extraction procedures and solvents used has led to the inconsistent lipid yields reported in the literature and industry, primarily because of the lack of a standard lipid quantification procedure, differences in compatibility of the polarity of the solvents chosen and the polarity of the lipid molecules present and accessibility of the lipids to solvent penetration. Inevitably, the extractable oil fraction will contain nonfuel components (e.g., chlorophyll, pigments, proteins, and hydrophobic carbohydrates), and thus it is necessary to assess the fuel fraction of these isolated oils (i.e., FA content of extracted lipids). This variability leads to the question of how one confirms actual improvements in particular strain

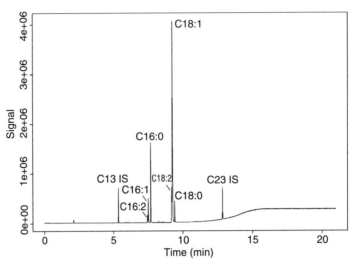

FIGURE 8.1. A typical FAME fingerprint chromatogram of *Chlorella vulgaris* UTEX395. The identification of the peaks is based on mass spectrometry (GC-MS) to identify the peaks. The designations C13 through C23 indicate the acyl chain length of the FAME; tridecanoate and tricosanoate methyl ester (C13 IS and C23 IS) were included as quantitative internal standards.

development programs. For example, is an observed 15% increase in extracted lipids translated in an increase in the fuel potential of the algae strain or an artifact of the measurement process? Is this difference smaller than the precision of the measurement methodology?

An alternative measure of the lipid content in algae is a whole biomass transesterification procedure. This method performs simultaneous hydrolysis and transmethylation of lipids in whole algal biomass and the resulting FA methyl esters are quantified by gas chromatography (GC; a typical chromatogram is shown in Figure 8.1). Since FAs make up the direct feedstock for lipid-based biofuels, an accurate measure of the total FA content is a better metric than through the process of lipid extraction. Of course, this ignores the constraints of the conversion process. Biodiesel production (based on alkaline $NaOH^-$ catalyzed transesterification of TAGs) has a low tolerance for polar lipids and free FAs. Catalytic hydrogenation to produce renewable diesel, jet fuel, or gasoline, has not been explored to that extent, and process constraints are not well established. Free FAs, in addition to TAGs, appear to be an acceptable feedstock. It seems likely that the refining industry, which has successfully addressed the challenges of changing petroleum feedstocks (light crude, high sulfur crude, heavy crudes, tar sands, etc.), will be able to also make use of polar lipids despite the inclusion of contaminating elements such as N, P, and S. The pivotal point in developing robust conversion processes will be the demonstrated ability to produce algal lipids in quantities consistent with the refining industry scale.

The use of an *in situ* transesterification procedure is already being used in the algae research community (13–14); however, the choice of catalyst varies and detailed description of the methodology is often lacking. These issues may hinder the adoption of this method as a standard procedure. The methods published typically have used a two-stage alkaline hydrolysis with NaOMe followed by an acid (BF_3) transmethylation of the fatty acyl chains and detection by GC; however, a detailed study of the parameters influencing the conversion efficiency was not reported. Recent work has yielded a simple one-stage acid hydrolysis (HCl) method for *in situ* transesterification and reports on a detailed description of the method, its influencing parameters and a direct comparison with the two-stage NaOMe:BF_3 procedure (15). This procedure was demonstrated to be robust across species; it is reproducible (with less than 3% relative variation between replicate measurements); and its efficacy is not dependent on the parameters listed above that influence an extraction process. This method can be adopted across algal strains, requires a small amount of biomass (4–7 mg, achievable in shake flask cultures), is accurate, reproducible, and precise, and can be applied in a rapid high-throughput manner to a large number of strains.

There is a continuing demand for higher-throughput analysis methodologies (see Chapter 5) to support research efforts to engineer or select superior algal strains as improved bioenergy feedstocks. These research efforts often require screening a large number of strains to identify one that accumulates high levels of desirable triglyceride lipids. Traditional analysis using chromatography methods are currently the bottleneck in such screening efforts. The number of samples in a typical screen can exceed several hundred, and it is not feasible to generate this amount of data using traditional analytical methodologies. This means that rich sources of biodiversity such as those included in large culture collections may be leveraged for the selection of superior bioenergy feedstock strains.

Fluorescent lipophilic dyes, such as Nile Red and BODIPY (4,4-difluoro-1,3,5,7,8-pentamethyl-4-bora-3a,4a-diaza-s-indacene), can be used for lipid visualization and strain screening because of their selective affinity for neutral lipid droplets inside the cells (16). However, a major disadvantage of the dye-based assays is that they are affected by uneven dye uptake due to the inherent variability of different strains of algae and their cell wall composition, which can be affected by growth conditions (17).

One technology that is able to address the issue of a comprehensive screening of a large number of samples is vibrational spectroscopy (in particular infrared [IR] spectroscopy) and can be applied to monitor the biochemical composition of algae over time. IR spectroscopy measures the absorption of energy in the IR region of the spectrum by chemical bonds in molecules. Changes in mid-IR spectra for biomass harvested from cultures at different time during the growth are shown in Figure 8.3. Because of the broad overtones of IR spectra, particularly in the near-IR spectra, the quantification of constituents heavily relies on the use of chemometrics, that is, multivariate calibration models. The advantage of IR spectroscopy is its tolerance to variation in the samples; spectral absorbance due to nonlipid components of the biomass can be subtracted in multivariate calibration models. IR spectroscopy also requires minimal sample preparation, is nondestructive, and is relatively independent of the biomass matrix. Overall, IR spectroscopy can be applied as a fast, accurate, and nondestructive analytical method that requires only very small amounts of homogenized biomass (~10 mg) using a 96-well plate setup. Calibration models have been developed that can be used as rapid high-throughput methods for the estimation of algal lipid content (18). Using the IR spectroscopic methods, algal lipid content of almost any algal species can be measured in a matter of minutes rather than days. Calibration models have been generated for single species as well as for multiple species combined, where the infrared spectra are correlated with lipid content. The main challenge with IR-based prediction of lipids is to have a good quality calibration model, which in turn depends on robust chemical data of a large set of "calibration samples." These prediction models can then be used to predict the lipid content in new, unknown samples.

8.1.3 Selection and Characterization of a Promising *C. vulgaris* Strain

The tools described above were applied to a set of 10 *C. vulgaris* strains available from the University of Texas (UTEX) culture collection in order to rapidly select and develop a high lipid-producing new algae lab strain. In order to move forward with a strain that has desirable parameters as a model production strain, we measured the growth rate and the lipid content of all 10 strains, grown under synchronized conditions. In order to rank the strains with regard to lipid and biomass productivity, growth experiments were set up in both nutrient replete and deplete media (in this case deplete refers to the lack of nitrate in the growth media). The biomass was collected from cultures after 10 days of growth in replete [+N] media, followed by 5 days of growth in deplete [-N] media. The growth of the cultures was measured as volumetric cell number since this is the most accurate measure of growth compared with measuring optical density. Lipid content is still expressed on a biomass basis, which still reflects the ultimate measure of productivity of a production system, and the data can feed directly into existing techno-economic models. For lipid quantification, the *in situ* transesterification method was used to avoid the inherent inaccuracies of gravimetric analysis of extracted lipids (15). Of the strains investigated, the UTEX395 *C. vulgaris* outperformed its competitors' growth rate and lipid productivity. The lipid content showed the biggest difference between the replete and deplete conditions: a 3.5-fold increase compared with a 1.7–2.3-fold increase seen for the other *C. vulgaris* strains. These data allowed us to conclude that UTEX395 had the most potential to be a production-relevant model organism; high growth rates corresponded with high lipid content and a potential to engineer metabolism to take advantage

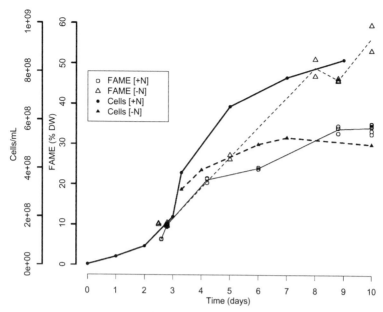

FIGURE 8.2. Growth of *C. vulgaris* (as measured cell density per mL of culture, closed symbols) and lipid content (FAME %dw as open symbols) over time in both a nitrate-replete (circles connected by solid line) and a nitrate-deplete (triangles connected by dashed line) culturing conditions.

of the big lipid content increases observed between replete and deplete growth conditions. This strain was then selected to move forward for further detailed investigation of the lipid productivity over the growth period. A study of the lipid content during the growth of UTEX395 under both replete and deplete conditions is shown in Figure 8.2. The data indicate exponential growth in replete media with a relatively low lipid content (<10%) until day 5, where the growth slows down and the lipid content of the harvested biomass increases to 20% and ultimately the lipid content after 10 days in replete media increases up to 35%. In deplete media the growth rate of the culture drops off dramatically at day 3 and the lipid content increases rapidly over subsequent days to close to 60% of the biomass dry weight (Figure 8.2).

The changes observed in overall lipid content over the growth of a culture reflect a significant shift in the biochemical composition of the biomass. In addition to building calibration models with IR spectroscopy, one can also observe overall biochemical changes in the biomass composition. Figure 8.3 illustrates the changes observed in mid-IR spectra over the course of the growth of a culture. Regions of the spectrum corresponding to carbohydrates, lipids, and proteins are highlighted and indicate considerable changes over the growth of the culture. The information present in the spectra can be used for building accurate calibration models to rapidly predict the concentration of

COMPLEX SYSTEM ENGINEERING

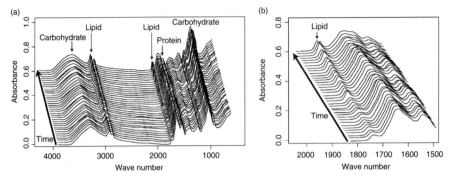

FIGURE 8.3. Illustration of changing mid-infrared fingerprints of algal biomass over the course of 3 weeks of growth and 2 weeks of nitrogen starvation (for lipid induction) (a). The region corresponding to lipids is shown in close-up, indicating significant increases over time (b).

lipids in algal biomass. This technology has the potential to rapidly increase the throughput of analyses and strain development discussed later.

8.2 DEVELOPMENT OF HYPOTHESIS-DRIVEN STRAIN IMPROVEMENT STRATEGIES

8.2.1 Systems Biology Analysis in an Unsequenced Microalga

Following downselection to a single algal cultivar (UTEX395), we next sought to utilize systems biology approaches to identify targets for strain improvement strategies aimed at optimization of lipid productivity. Systems biology, or "omics" analyses (see Chapter 3), such as comparative transcriptomics (RNA-seq) and proteomics, offer valuable platforms for the development of hypothesis-driven metabolic engineering strategies through the elucidation of key biosynthetic components involved in algal lipid production. However, these analyses are highly dependent on available genomic sequence data. Obtaining such genomic sequence data can often be cost-prohibitive, and efficient assembly and bioinformatics analysis of such data can be extremely time and labor-intensive. As such, many promising strains of potential commercial-relevance remain relatively unexplored, with most systems analysis to date largely focused upon established model organisms (19). While such model systems offer valuable insight into general algal biology, they may fail to present appropriate models for elucidation of the molecular underpinnings of high lipid productivity in oleaginous microalgae. For example, the model organism *C. reinhardtii*, which has served as the platform for the majority of fundamental microalgal research due to its well-established laboratory cultivation and genetic transformation systems, only produces ~20% lipid on a dry-cell weight (dcw) basis under nitrogen deprivation (13). By comparison, the unsequenced, oleaginous microalga *C. vulgaris* (UTEX395) produces ~60% dcw lipid under similar wild-type conditions (Figure 8.2).

As discussed in Chapter 3, transcriptomic analysis allows for expression profiling of mRNAs present in a given cell population under varying growth conditions. Comparative RNA-seq analyses under varying growth conditions can thus implicate genes and gene sets (gene set enrichment analysis) responsible for phenotypes of interest. In this case study, the phenotype of interest is the high lipid accumulation observed in *C. vulgaris* UTEX395 under nitrogen limitation. Although traditional transcriptome assembly is performed through mapping of short sequence reads to a complete genome, emerging technologies now allow for *de novo* assembly of short cDNA sequence reads in the absence of available genome information. We took advantage of these technologies, utilizing the Velvet and Oases software packages to assemble the short reads obtained via Illumina sequencing (20,21). Assembly of a *C. vulgaris* transcriptome will ultimately allow for mRNA expression profiling, although methods for accurate transcript quantitation in the absence of a genome require further development (discussed below). More immediately, however, we sought to utilize the assembled transcriptome for a less obvious purpose, proteomic analysis.

8.2.2 Transcriptome-to-Proteome Pipelining

Although transcriptomic analysis offers a great deal of insight into the genetic control of product formation, it does not fully define this regulation, as mRNA expression levels are not always proportional to the expressional levels of protein for which they code. Additionally, higher plant and algal metabolic regulation has been shown to be largely influenced by post-transcriptional gene regulation (22–27). As such, proteomic analysis is a critical complementary tool in the successful strain engineering of a commercially relevant oleaginous microalga. However, proteomic analysis of unsequenced microalgae (and unsequenced organisms in general) is challenging, largely due to the low peptide identification rates associated with orthologous database searching (19). Proteomic analysis using orthologous sequence databases requires nearly identical mass/charge values (±1–2 Da) between the search database peptides and peptides of interest in order to match an equivalent mass/charge ratio of statistical significance and avoid the production of unmanageably large result databases of questionable reliability. As such, a single amino acid differential between a search model sequence and a peptide fragment sequence of interest can result in a failure to produce a statistically significant match, leaving significant gaps in protein identification (19). We bypassed the necessity for genomic sequence data by moving directly from the *de novo* transcriptomic assembly discussed above, to proteomic analysis using the resultant assembled transcriptome as a search database (Figure 8.4). Through six-frame translation, transcriptome sequence data can be matched to mass spectral peptide data, offering a powerful proteomic database. This strategy is

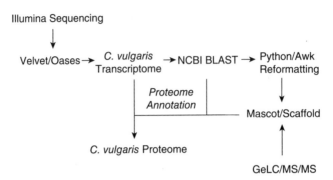

FIGURE 8.4. Workflow for *de novo* transcriptome assembly and comparative proteomic analyses in an unsequenced organism. Adapted from Reference 19.

advantageous in that it meets the stringent requirements for sequence accuracy of translated gene sequences needed for whole organism shotgun tandem mass spectrometry (MS/MS) approaches. Additionally, utilization of a transcriptome-to-proteome pipeline allows the coupling of gene and protein annotation, again allowing for increased throughput in systems data analysis. As discussed below, the utilization of a transcriptome-to-proteome pipeline dramatically enhanced our proteomics results.

We utilized the lipid content of our chosen model strain as the basis for comparative proteomic analysis. Lipid accumulation throughout the growth cycle was examined using the FAME analysis tools described above (Figure 8.2), and was utilized to select optimal harvest points for comparative proteomics. Cells were harvested under both nitrogen replete (low lipid) and nitrogen deplete (high lipid) conditions, and soluble protein fractions from whole cell lysates were obtained for samples corresponding to 10% and 60% FAME. Proteomic analysis was performed using gel-based liquid chromatography mass spectrometry (GeLC/MS). Product ion data were searched against both *Chlorophyta (all available green algal genome sequences, which at present consists of C. reinhardtii, Ostreococcus sp., Coccomyxa sp. C-169, Micromonas pusilla, Volvox carteri, and Chlorella sp. NC64A)* and a six-frame translated *de novo* assembled *C. vulgaris* transcriptome database. In order to utilize the *C. vulgaris* transcriptome in this capacity, we developed a pipeline of in-house Python and Awk scripts in order to annotate the transcriptome using basic local alignment search tool (BLAST) (blastn) results, and to properly format the resultant annotated transcriptome for use in Mascot, which allows for six-frame translation and database interrogation (28). Each transcript isoform in the assembled transcriptome was annotated using the fasta header of the best blastn hit, which in turn was utilized to annotate positive MS/MS peptide identifications. The numbers of proteins, matching spectra, unique peptides, mean and median spectra/protein, and mean and median unique peptides/

TABLE 8.1. Improved Database Interrogation Using a *De Novo* Assembled Transcriptome. Adapted from Reference 19

MS/MS Data Acquisition (Average for Both N-replete and N-deplete)	All *Chlorophyta* Genomes	*C. vulgaris* Transcriptome
No. of proteins	2,061	2,949
No. of matching spectra	25,638	64,923
No. of unique peptides	7,831	30,543
Mean spectra/protein	17	26.7
Median spectra/protein	9	16
Mean unique peptides/protein	5.2	12.6
Median unique peptides/protein	5	9

protein all increased approximately twofold using the *de novo* assembled *C. vulgaris* transcriptome compared with the *Chlorophyta* database, clearly indicative of a superior search database (Table 8.1).

Further underscoring the advantages of a transcriptome-to-proteome pipeline, the utilization of the *C. vulgaris* transcriptome database identified a number of proteins along the central metabolic pathways that were initially absent from the data obtained using only *Chlorophyta* sequence databases. Many of the proteins in this newly identified dataset play critical roles in FA and TAG biosynthesis. A schematic of the enzymatic components involved in FA and TAG biosynthesis is shown in Figure 8.5. Although orthologous searching identified only three enzymatic components of the FA biosynthetic pathway, and none of the TAG enzymatic components, utilization of the *C. vulgaris* transcriptome as a search database allowed us to identify the enzymatic components of the FA and TAG biosynthetic pathways in their entirety (Figure 8.5). We examined changes in spectral counts (indicative of protein abundance) for the components of the FA and TAG biosynthetic pathways under nitrogen-replete and nitrogen-deplete conditions. Components of the FA biosynthetic pathways demonstrated relatively minor changes in abundance (approximately one- to twofold) (Figure 8.5). Conversely, the change in protein abundance for the TAG biosynthetic pathway was far more pronounced, one to two orders of magnitude greater than those observed in the FA biosynthetic pathway (Figure 8.5).

Our results demonstrate the necessity for accurate sequence information in proteomic analysis, and more importantly, the utility of a *de novo* assembled transcriptome as a search model for proteomic analysis of unsequenced microalgae. Bypassing the necessity for genomic sequence data avoids the time- and cost-prohibitive nature of complete genome assembly and annotation. A transcriptome-to-proteome pipeline narrows sequence data down to just coding sequences, avoiding intronic regions and thus allowing for more rapid assembly and interpretation of sequence data using readily available RNA-seq

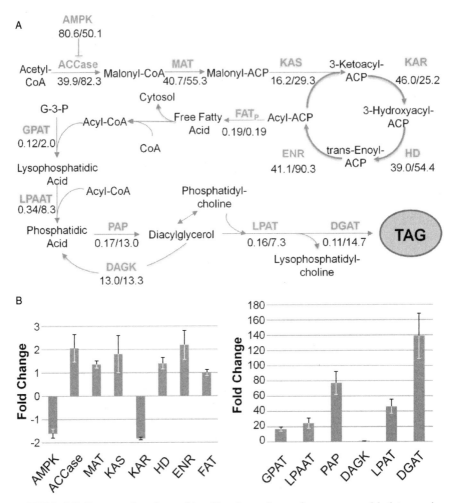

FIGURE 8.5. Improved pathway identification using a *de novo* assembled transcriptome database and changes in protein abundance under nitrogen depletion. (A) Critical components of the fatty acid and triacylglycerol (TAG) biosynthetic pathways were absent from initial MS/MS searches against all available *Chlorophyta* databases. All proteins were absent from initial MS/MS identification, except for AMPK, ACCase, and ENR, yet positively identified when searching against the *C. vulgaris* transcriptome database. Numbers below proteins represent normalized spectral abundance factor (NSAF) values (10^5) for nitrogen-replete and nitrogen-deplete conditions, respectively. ACCase, acetyl-CoA carboxylase; ACP, acyl carrier protein; AMPK, AMP-activated kinase; DAGK, diacylglycerol kinase; DGAT, diacylglycerol acyltransferase; DHAP, dihydroxyacetone phosphate; ENR, enoyl-ACP reductase; FAT$_P$, fatty acyl-ACP thioesterase (putative); G3PDH, glycerol-3-phosphate dehydrogenase; GPAT, glycerol-3-phosphate acyltransferase; HD, 3-hydroxyacyl-ACP dehydratase; KAR, 3-ketoacyl-ACP reductase; KAS, 3-ketoacyl-ACP synthase; LPAAT, lyso-phosphatidic acid acyltransferase; LPAT, lyso-phosphatidylcholine acyltransferase; MAT, malonyl-CoA:ACP transacylase; PAP, phosphatidic acid phosphatase. (B) Corresponding spectral count fold-changes for components of the FA (left panel) and TAG (right panel) biosynthetic components. Reproduced from Reference 19.

software. In turn, it provides a platform on which to perform systems biology analyses in unsequenced, "nonmodel" organisms, leading to the identification of strain engineering targets in organisms of commercial relevance.

Although transcriptome-to-proteome pipelining offers a number of advantages, it does have limitations. Incomplete sequence data potentially limits the identification of promoters, coding start and stop sites, and stretches of internal coding sequences (largely an artifact of current NextGen sequencing platforms), all of which are critical for the development of transformation methods (discussed in the following section). In addition, incompletely assembled transcripts (an artifact of current assembler tools) constrain annotation, limiting positive protein matches. Likewise, transcripts that are temporally expressed may not be identified under certain harvest conditions, and in turn will not yield a positive protein hit. Small regulatory RNAs (such as micro-RNAs) may also be absent under certain harvest conditions, and are frequently discarded by currently available *de novo* assembler programs. However, it is worth noting that with the rapid pace at which next generation sequencing tools are advancing, these limitations are likely to be minimized soon. Regardless, transcriptome-to-proteome pipelining offers a rapid, effective tool for identifying strain engineering targets in organisms for which a genome sequence is lacking.

8.2.3 Identification of Strain Engineering Targets

The dramatic protein abundance differential between FA biosynthetic and TAG biosynthetic components may imply that TAG biosynthesis plays a significant role in the rate-limiting production of neutral lipids, suggesting that future studies aimed at strain improvement might be focused on overexpression of TAG biosynthetic components. Although all proteins in this pathway were greatly increased in abundance following nitrogen starvation, the largest increase was observed for diacylglycerol acyltransferase (DGAT), the enzyme responsible for committed entry into TAG biosynthesis, with greater than 100-fold spectral count increase, making this an attractive target for engineering strategies (Figure 8.5). Indeed, overexpression of DGAT in higher plants has already been shown to increase TAG accumulation, indirectly validating this hypothesis (29,30).

Committed entry into FA biosynthesis is also of interest, as it may be an upstream bottleneck of neutral lipid synthesis. acetyl-CoA carboxylase (ACCase) governs entry into FA biosynthesis, and early algal strain engineering strategies from the ASP targeted this enzyme for expression (although no increase in lipid accumulation was observed). Interestingly, AMPK, an ACCase inhibitor, was downregulated under high lipid-producing conditions (Figure 8.5). This lends potential insight into the regulation of FA synthesis through rate-limiting ACCase activity. It is possible that AMPK plays a critical role in driving the equilibrium between acetyl-CoA and malonyl-CoA in the reverse direction, ultimately slowing the rate of FA biosynthesis and increasing the rates of FA beta-oxidation. The activity of AMPK under nitrogen-replete and nitrogen-deplete conditions warrants further investigation, and also presents a potential strain engineering target. FA and TAG targets, however, represent just some of the many proteins identified with extreme differential abundance. As data analysis progresses, we will generate a number of additional targets from less obvious pathways.

Beyond target identification, utilization of the *C. vulgaris* transcriptome also allowed for identification and differentiation of protein isoforms. Homomeric and heteromeric ACCase isoforms, as well as multiple ketoacyl-ACP synthase (KAS) isoforms, were identified during the annotation stage. Isoform differentiation can have a dramatic impact on strain engineering strategies. For example, it has been suggested that overexpression of cytosolic homomeric ACCase, coupled with plastidial sub-cellular localization, as opposed to overexpression of the more complex, multi-subunit heteromeric plastidial isoform, may be a simpler and more efficient means to increase FA content in oleaginous organisms (31). Targeted strain improvement efforts and complete pathway analyses will thus be greatly facilitated by the isoform identification and maximal identification coverage that a *de novo* assembled transcriptome search database affords.

We have focused our initial investigation of differential protein expression upon dramatically different lipid accumulation states (see Figure 8.2) in N-replete and deplete *C. vulgaris*. These analyses indicate that the FA and TAG biosynthetic pathways are upregulated under nitrogen limitation, especially for the case of TAG components. We hypothesize that future analyses using intermediate harvest points will lead to a less pronounced differential between FA and TAG biosynthetic components, with an increased abundance of FA components and a decreased abundance of TAG components prior to nitrogen exhaustion. Future analyses will therefore be focused on intermediate time points for accumulation, which will allow for abundance mapping throughout the lipid accumulation cycle and help clarify the *rates* of TAG biosynthetic component expression. Concurrently, quantitative analyses of TAG and TAG biosynthetic intermediates, such as phosphatidic acid and diacylglycerides, will lend further insight into the flux through the TAG pathway, as well as temporal regulation throughout the lipid accumulation cycle. Data from intermediate accumulation states will also likely provide a wealth of additional information with regard to the stages at which gene and protein expression are initiated. However, a more complete, integrated systems biology analysis, incorporating transcriptomic, proteomic, and metabolomic data will be necessary to fully elucidate potential flux bottlenecks in the FA and TAG pathways. At present the most effective means of quantifying RNA-seq data in the absence of a genome is analyzing reads per kilobase of exon model per million mapped reads (RPKM). Efforts are currently under way to improve this methodology, as expression profiling will be an essential component in downselection of targets. As mentioned above, we have focused our initial investigation on the most obvious pathways of interest, namely the FA and

TAG biosynthesis pathways. However, initial results also suggest that dramatic changes in protein abundance are occurring in other central metabolic pathways, transcription factors, lipases, translation machinery, and many other factors, all of which may ultimately prove to be key targets for strain manipulation. Iterative compositional analysis, using the near-infrared (NIR) methods discussed in Section 8.1.2, will allow for high-throughput assessment and validation of genetic manipulation of these targets (discussed below) and resultant product formation.

8.3 IMPLEMENTATION OF BIOLOGICAL TOOLS I—DEVELOPMENT OF A TRANSFORMATION SYSTEM

With a number of promising strain improvement targets identified, we next sought to develop an effective transformation system to initiate our hypothesis-driven strain improvement strategies. As discussed above, meeting the economic goals of biodiesel production requires optimization of many complex phenotypes in algal strains, including growth rates, lipid production and accumulation capacity, and contamination control. Genetic engineering offers not only the means to manipulate a strain by introducing, removing, or modifying DNA, which results in new or more desired phenotypes, but also the means to test or confirm a hypothesis derived from advanced genomic studies, such as transcriptomics or proteomics. A tremendous amount of data have been generated using transcriptomics and proteomics on our model microalga, *C. vulgaris* UTEX395, in our laboratory. Expression of numerous genes in the lipid and FA pathways was found to be significantly upregulated or downregulated under nitrate starvation (19). This information provides valuable insight into the regulation mechanisms of lipid synthesis, and these genes can be further used as targets for modification to improve the pathways. In any case, capability to transform *C. vulgaris* is a crucial part of genetic engineering in our microalgae projects. Although several publications have reported successful genetic transformation in *Chlorella* spp. (32–40), these methods tend to be strain specific and in many cases the results are not reproducible. In order to fully utilize the information obtained from our transcriptomic and proteomic studies and to modify our model strain, *C. vulgaris* UTEX395, for desired characteristics using genetic engineering, we initiated the development of an efficient transformation system in this microalga.

8.3.1 Vector Construction

To assemble a vector for efficient transformation, several elements need to be considered. Typically, antibiotic resistance is used as selection for the transformants. We have tested the sensitivity of UTEX395 to a few antibiotics. The results indicated that UTEX395 is highly sensitive to phleomycin, zeocin, and G418, and moderately sensitive to paromomycin and hygromycin. Therefore, genes conferring resistance to the mentioned antibiotics (*sh ble*, *nptII*, *aphVIII*, and *hph*) may be incorporated in the vectors. Promoters are important elements for vector construction and a subject area that will be discussed further below. In the initial vector construction for transformation development, we used a *Chlorella* viral promoter, AMTp (adenine methyltransferase (41)) for the expression of *ble* and the promoter of *C. vulgaris* nitrate reductase gene for the expression of *egfp* (42). Terminator sequences used were of HSP70 (heat shock protein) and nitrate reductase genes, respectively, from *C. vulgaris*. It has been noted that codon optimization can be prudent for the expression of genes in some algae (43,44). We have taken this into account and optimized the genes of interest (*ble* and *egfp*) based on the known *Chlorella* codon usage table when constructing our vectors.

8.3.2 Protoplast Preparation and Transformation of *C. vulgaris* UTEX395

A variety of methods have been employed for *Chlorella* transformation. These methods include electroporation, microprojectile bombardment, glass bead agitation, and protoplast transformation (45). One of the major barriers to transformation is the resilient cell wall structure of *Chlorella*, which makes DNA penetration into the cells more difficult (45,46). To overcome this challenge, much effort was undertaken for the protoplast formation in UTEX395 to ensure the uptake of DNA. Protoplast generation was achieved by treating the cells with cell wall-degrading enzymes, cellulase, acromopeptidase, and macerozyme (47). Multiple transformation protocols (35,39,48,49) were adopted and modified into the current protocol for UTEX395. Enzyme-treated protoplasts of UTEX395 were incubated with the vector DNA in polyethylene glycol (PEG) and lithium acetate followed by addition of dimethyl sulfoxide (DMSO) and a heat shock at 42°C. Transformed cells were mixed with soft agar (0.75%) and plated on agar medium plates containing the antibiotic phleomycin. Treatment with cell wall-degrading enzymes followed by heat shock enhanced the transformation efficiency by at least 10-fold relative to the untreated, intact cells (Figure 8.6A). Successful transformation was confirmed by polymerase chain reaction (PCR) using genomic DNA (gDNA) of the transformants and primers specific to *ble* (Figure 8.6B). Furthermore, expression of *ble* in the transformants was demonstrated by quantitative real-time PCR (qRT-PCR) (data not shown). Transformants exhibited a range of *ble* expression levels up to threefold difference. However, despite the phenotypic and genetic evidence for the transformants obtained, we have subsequently experienced challenges in the reproducibility of the transformation results. Several protoplast transformation experiments were attempted after the initial successful transformations; however, to date, we have been unable to generate new transformants using the protocol described above. This phenomenon is not uncommon in the area of algal transformation, and many factors may be attributed to this irreproducibility (50,51). For example, the importance of endogenous promoters for heterologous gene expression was

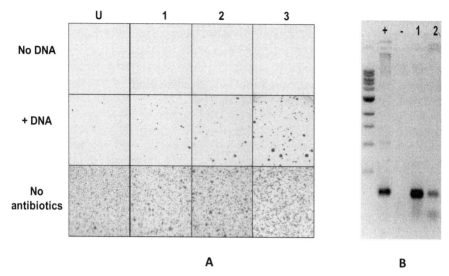

FIGURE 8.6. (A) Transformants of *Chlorella vulgaris*. U: untreated cells, 1–3: cell wall-degrading enzyme-treated cells. (B) PCR analysis of transformants. +: Vector DNA; –: untransformed gDNA; 1, 2: transformants gDNA. *(See insert for color representation of the figure.)*

recently demonstrated in the oleaginous alga *Nannochloropsis gaditana* (51). We have taken into account this understanding, and redesigned our plasmids to incorporate strong, endogenous promoters from *C. vulgaris* UTEX395 to drive *ble* expression. We are currently evaluating these vectors using the methodology described above.

8.3.3 Stability Evaluation of Transformants

A second major barrier to a successful transformation is the maintenance of the foreign DNA in the genome of the host. Transient expression and instability have been observed in *Chlorella* transformation (33,52), although claims were also made with stable transformation and expression of certain foreign proteins (35,37). The instability is indicated by the gradual loss of phenotypes during subculturing of transformants in the absence of selection pressure. Our group views the stability of transformants as an important criterion for a useful transformation. Although we noted above our uncertainty regarding the status of the transformants generated by protoplast/PEG transformation, we believe that the subsequent steps taken to test their stability are worth describing to serve as a guide for others following this path. To test stability, 25 phleomycin-resistant transformants were cultured in the growth medium without phleomycin for 10 successive transfers followed by scoring the phleomycin-resistant population from each transfer. The results indicated that over 60% of the transformants were stable for at least 10 transfers (30 generations) in the absence of the selection pressure, phleomycin. Less than 40% of the transformants exhibited poor stability. More work is ongoing to study the instability of this population.

8.3.4   *C. vulgaris* **Endogenous Promoter Identification and Characterization**

Establishment of a stable transformation system is a prerequisite for a useful genetic tool box. In addition, among the many regulatory factors in microalgae, promoters may play an important role in modulating the expression level of genes. To modify the metabolic pathways in microalgae using genetic engineering, availability of inducible promoters or those with different strengths will add a great attribute to the tool box. Promoters from a variety of organisms, including mammalian cells, diatoms, *Chlorella* spp., plants, and viruses have been demonstrated to be functional for gene expression in *Chlorella* spp. (32,36,38,39,53). Among these studies, viral promoters (e.g., CaMV 35S) appear to be the most commonly used, with the exception of nitrate reductase promoter from *C. vulgaris* (36). Due to the complex genetic diversity of *Chlorella* spp., we thought that transformation and gene expression in *C. vulgaris* UTEX395 would benefit from the use of the endogenous promoters, as discussed above. Taking advantage of the systems biology data (a combination of transcriptomics and proteomics analyses, discussed above) from UTEX395, we have identified several genes in the lipid and FA pathways that have elevated or reduced expression due to nitrate starvation. Promoters of those genes will be isolated by genome walking techniques (54) and characterized using a reporter gene, such as *egfp* (42) or *uidA* (55), for their strength. Similarly, constitutive promoters (strong or weak), indicated by the expression level of genes, will be of great value to us in the strain improvement of *C. vulgaris* and are in the plan for isolation and further characterization.

8.4   IMPLEMENTATION OF BIOLOGICAL TOOLS II— DEVELOPMENT OF A SELF-LYSING, OIL-PRODUCING ALGA FOR BIOFUELS PRODUCTION

8.4.1   Algal Lipid Extraction

Although the majority of the work presented thus far has focused on "upstream" strain improvement strategies, namely strategies to increase oil accumulation, it was also important to think ahead to downstream strain engineering strategies focused on product isolation and recovery. Even with highly productive strains, many challenges remain in the development of an algal biomass to liquid transportation fuel process (56–58). Among these is the challenge of extracting the internal oil stores for processing into finished biofuels. Few other problems have generated as much interest from the industrial and research communities, and numerous approaches have been explored. Many of these methods rely on solvent extraction from intact or mechanically ruptured algal cells and, depending on the solvent used, may also require extensive drying, a potentially huge cost driver. Solvents such as toluene, hexane, butanol, ethanol, methanol, and ionic liquids are being considered (59–62). Depending on a number of factors, solvent extraction may not be optimal as microalgae are known to have thick, complex, recalcitrant cell walls. Solvents must pass through the cell wall, cell membrane, and oil body membrane to interact with the internal algal oils and then reverse this movement to transport the oils outside the cell. In our techno-economic model, we have included a mechanical disruption step to facilitate extraction, although it adds both cost and energy demands to the overall process (9). Additionally, solvents are usually classified as hazardous materials, are expensive, and would require complete recovery and recycling to be economically viable. Finally, solvent recycle is poorly understood and life cycle analysis suggests that a significant amount of solvent, up to 2 g of hexane per kg of biomass treated, will be lost during processing (63).

In an effort to avoid using solvents, alternative methods are being pursued that rely on external energy inputs in the form of ultrasound, electromagnetic pulses, and physical disruption, or on chemical acid or base treatments (59,64,65) to either augment or replace extraction. These methods may be costly due to the high energy required to rupture the algal cell walls. In techno-economic studies, the extraction process has been modeled to be one of the top contributors to both the capital equipment costs and operating costs of an algal biorefinery process (9,62,66). Therefore, it represents a key opportunity for cost reduction. Our entry point into this challenge is to look to the natural processes and to take advantage of the biology. Biological systems, having been tuned to efficiently overcome entropy over evolutionary time, are typically simple, specific, and, most importantly, function with the lowest energy input required. Thus, an elegant solution with low energy and chemical inputs, exemplified by secretion in current fermentation processes, would take advantage of a natural, inducible cellular response. This would require development of an industrially relevant, oil-producing algal strain with a complex phenotype. Ideally, this means an algal strain capable of high oil production with controlled, self-induced cell wall degradation that releases internal organelles, oil bodies, under a controlled external stimulus. Our strategy to achieve this goal has been to develop an enzyme-based process to facilitate oil release (Figure 8.7). An additional benefit is that enzymatic treatment of algal biomass would leave the residual algal biomass pretreated in a way that downstream processes such as nutrient recycling, anaerobic digestion, thermal depolymerization, or gassification may be more facile. Enzymatic degradation has the potential to vastly simplify the harvesting, dewatering, and oil extraction processes. We envision a process where algae will be partially dewatered, perhaps to 20% solids, then induced for self-lysis by partial cell wall degradation. Oil bodies will escape from the cells and can be easily recovered by simply skimming the surface, or using an established emulsion breaking process, or using a recycled portion of the algal oil stream for enhanced recovery. To develop

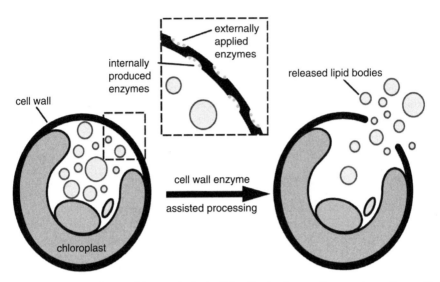

FIGURE 8.7. Release of internal algal oil bodies by internally or externally applied enzymes. *(See insert for color representation of the figure.)*

an industrial algal strain for use in a process involving an enzymatic degradation step, a suite of enzymes must be tested for efficient cell wall degradation. The relationship of cell wall degradation to the release of oil bodies must be determined. Finally, production of the enzymes will need to be established in the algal host under inducible promoter control that does not require expensive reagents or conditions to effect an economically viable induction of the enzymatic degradation and subsequent oil release.

8.4.2 Algal Cell Wall Complexity and Enzymatic Treatment Effects

Due to the chemical complexity of the polymers involved, and the structural complexity of cell wall architecture, there is still a considerable knowledge gap concerning accurate and definitive algal cell wall composition for most candidate species (67,68). Our approach to deconstructing algal cell walls has been to utilize digestive enzymes themselves to derive information about specific linkages present in algal cell walls and how those linkages can be exploited to promote oil body release. This approach will not necessarily fill in the knowledge gap concerning specific algal cell wall polymers, but enzymes can help determine cell wall structure and composition by providing details about specific glycosidic linkages present. Information gained in this way can then be used to figure out how to best break down algal cell walls. A two-pronged strategy was employed to find effective enzymes: examining the impacts on colony growth and the impacts on mature cells by tracking increasing permeabilization via the entry of a DNA staining dye. An enzyme impacting growth is important during formation of the cell wall and if it inhibits growth, one can presume that

TABLE 8.2. Growth Inhibition of *C. vulgaris* by Various Enzyme Classes

Enzyme	Inhibition
Alginate lyase	No
β-glucuronidase	++
Cellulase	No
Chitinase	+++
Chitosanase	+
Dreiselase	No
Hemicellulase	No
Hyaluronidase	No
Lysozyme	+++
Lyticase	No
Macerozyme	No
Pectinase	++
Pectolyase	No
Sulfatase	++
Trypsin	+
Xylanase	No
Zymolyase	No

by preventing those specific linkages from forming, the enzyme is preventing a mature cell wall from being established and thus those susceptible linkages and components are present. For mature cell walls these same enzymes may or may not still work because the target glycosidic bonds may now be inaccessible in the complex architecture of the mature cell wall. The linkages that were available to enzymes as the cell wall polymers were first produced may now be enclosed in a matrix of other materials or buried deep within the cell wall. We used a plate-based assay to determine the effects of various enzymes from different classes on the growth of *C. vulgaris* UTEX395. By inoculating a dilute culture into appropriate nutrient containing soft top-agar and then spotting enzymes directly on this top-agar, while the dilute culture is growing, zones of inhibition will appear around active enzymes. Table 8.2 describes the effects of various enzymes having different classes of enzymatic activity.

Several enzymes, chitinase, sulfatase, β-glucuronidase, pectinase, and lysozyme strongly inhibit growth of *C. vulgaris*. Cellulase, hemicellulase, and xylanase do not inhibit growth, suggesting a lack of accessible cellulose or hemicelluloses such as found in higher plant cell walls. Alginate lyase which cleaves β-1-4 mannuronic bonds, also showed no inhibition of growth. We analyzed some of the more effective growth inhibiting enzymes both singly and in combination with lysozyme for their effects on mature, nitrogen-sufficient cells in overnight digestions. The cells were then incubated with a DNA staining dye, SYTOX green, which only stains compromised, permeable cells. In the absence of enzymes, cells were not permeable to the dye and after exposure to various enzymatic activities a portion of the population became permeable (Table 8.3).

TABLE 8.3. Percentage of *C. vulgaris* Population Permeable to SYTOX Green Dye after Enzymatic Treatment

Enzyme	% Permeable	% Permeable + Lysozyme
No enzyme	2.2	—
β-glucuronidase	2.6	54.1
Cellulase	1.2	21.1
Lysozyme	11.9	—
Lyticase	1.09	48.4
Pectinase	1.45	32.7
Sulfatase	1.5	98.8
Trypsin	0.9	29.9

The results of the cell permeabilization experiments suggest that a coating of chitodextrin (β-1-4 linked N-acetylglucosamine) or peptidoglycan (β-1-4 linked N-acetylmuramic acid and N-acetylglucosamine)-type material, both polymers sensitive to lysozyme, surrounds or otherwise protects many of the other polymers from enzymatic attack. It is only after lysozyme strips away or damages the outer layer that other enzymes are then able to act on the cell wall, causing increased permeabilization. In some cases the results are dramatic. Treating *C. vulgaris* with lysozyme and sulfatase permeabilizes nearly 100% of the cells, whereas with lysozyme alone, only 12% of the population is permeabilized. Sulfatases hydrolyse O- and N-linked sulfate ester bonds, suggesting that sulfated polymers are integral to cell wall architecture in *C. vulgaris*. It is also interesting that some enzymes have a large effect on growing cells by inhibiting growth yet do not seem to have much effect on permeabilizing the cell walls of mature cells. As an example, cellulase and lyticase applied individually do not have much effect on growth. However, each in combination with lysozyme permeabilizes up to 21% and 48% of the *C. vulgaris* population, respectively. These results suggest that algal cell wall sensitivities to enzymatic activities change as the cell matures. Perhaps cellulose is synthesized in the cell wall at a later maturing stage and is covered by another resistant polymer during this process such that cellulase cannot inhibit growth and can only attack the embedded cell wall cellulose once it is rendered accessible by another enzymatic activity such as lysozyme.

8.4.3 High-Resolution Imaging of Enzymatic Treatment Effects

To further explore the nanoscale architecture of the algal cell wall that may underlie these results from enzyme mixture digestions, we employed surface characterization by high-resolution imaging. Transmission and scanning electron microscopy were used to directly visualize the effects of enzymes on algal cell walls. The effects of lysozyme on the cell walls of *C. vulgaris* are dramatic and complex. Transmission electron micrographs reveal the complete loss of the hair-like fiber layer of the outer wall surface, swelling of the outer layers, and a peeling or dissolution of material from the outer cell wall (Figure

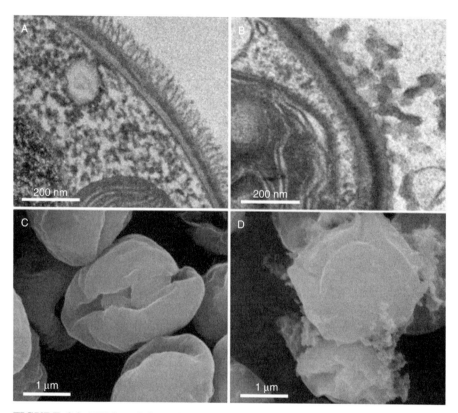

FIGURE 8.8. TEM and SEM images showing changes in and degradation of *C. vulgaris* cell walls by lysozyme.

8.8A,B). At first glance, it seems counterintuitive that a digested cell wall that has lost material would appear thicker. In fact, this is typical for a complex, compact, layered cell wall to swell significantly as its internal cross-linked structure is weakened. Although there is no apparent pitting or other surface defects readily observable by scanning electron microscopy (SEM), the same amorphous extracellular matrix from degradation of the cell wall is clearly apparent (Figure 8.8C,D). This extracellular material appears to derive from dislodged outer cell wall layers and is still attached to the cell by fibrous strands. More work remains to determine what the key architectural changes are and the level of structural disruption that will be required to generate a self-lysing phenotype. It is almost certain that the cell wall does not need to be entirely digested away to improve oil extraction; however, it may need to be permeabilized beyond what is required for a small dye molecule to pass through.

8.4.4   Production Strain Development

As mentioned previously, the growth assays, permeabilization, and surface characterization studies do not provide an unambiguous determination of the composition of algal cell walls but do provide the critical information on the types of linkages present and indicate how to functionally degrade the algal cell walls. Using the data from these experiments, a cocktail of enzymatic activities for efficient cell wall disruption can be created either from enzymes in-hand or through the mining of transcriptomic and proteomic datasets to provide sequence data on native enzymes possessing the desired enzymatic activity. Native, intracellular cell wall-degrading enzymes needed for cell division to partially degrade the algal cell wall have been described (69–71). The enzyme screening experiments demonstrated that multiple layers of differing cell wall material will need to be degraded. This will require a combination of synergistic enzymatic activities. The data suggest that 2–4 different enzymatic activities should be sufficient to penetrate or weaken the cell wall sufficiently to enhance lipid extraction. Engineering an algal strain to produce a small number of additional enzymes will likely not pose much of a metabolic burden.

The final step in effecting an elegant solution to this strategy is the development of the production organism. This involves, as discussed above, the tightly controlled induction of the relevant enzymes. The genes of interest will have to be placed under the appropriate expression controls and stably transformed into the host organism. Transformation of many walled organisms requires some level of cell wall permeabilization in order for intact DNA to pass through to the nucleus (72). In algae this may be particularly important, as algal cell walls are known to be tough and resistant to a wide variety of stresses. Additionally, to effectively express cell wall-degrading enzymes in a green alga such as *C. vulgaris*, native expression systems will be required. Of critical needs are those that are tightly regulated and have a rapid, specific, and effective signal to induce high levels of expression. Inducible promoters responding to changes in pH or temperature may be useful but ultimately not specific enough. Although engineering controls can be very effective at maintaining well-mixed growth conditions, numerous micro-environments will still exist in the bends, eddies, CO_2 sumps, and other incongruous areas of large open ponds. Thus, pH and temperature signals may not be specific enough, and addition of an inducing chemical and requisite genetic control will be required.

In conclusion, the solutions to difficult hurdles in biological processes, such as algae-to-liquid transportation fuels, often require the development of complex biological phenotypes and rely on pursuing multiple strategies concurrently. In the case of the construction of a self-lysing, industrially robust, oil-producing algal strain, this is certainly true. Green algae do not enjoy the same well-established genetic engineering tools and methods as other organisms, such as *E. coli* and yeast, yet through focused persistent efforts, success can be achieved. As one project seeks to develop reliable transformation strategies in green alga, discussed in Section 8.3.2, this project seeks to identify the appropriate genes to then utilize these systems to achieve internal, tightly controlled expression of cell wall-degrading enzymes.

8.5 CONCLUDING REMARKS

Pursuing a hypothesis-driven strain-improvement program in an unsequenced microorganism can present a number of unique challenges. Successfully implementing genetic engineering strategies in such an organism requires a multifaceted, yet integrated effort. In the current case study we have presented an efficient strategy for the initiation of such efforts. Careful selection of a commercially relevant microorganism is a critical first step. The development of compositional analysis tools for selection of ideal strains and validation of downstream strain engineering strategies thus served as a platform on which to build our program. Once a promising strain was selected, the development of a strain engineering toolbox must be rapidly developed and implemented. We have demonstrated that an omics pipeline focused on transcriptome-to-proteome analyses can be applied to generate promising targets for genetic and metabolic engineering. We then presented an ongoing strategy for the development of a transformation system with which to genetically manipulate the targets identified through omics analyses or implement downstream targets focused on improved product recovery. These strain development strategies are intimately linked, and successful complex phenotype engineering will ultimately rely on utilizing these processes iteratively.

ACKNOWLEDGMENTS

The authors are grateful to Al Darzins for technical leadership on several projects described in this chapter, Henri Gerken for enzyme screening and imaging work, Todd Vinzant for assistance with algal imaging at NREL's Biomass Surface Characterization Laboratory, Hua Zhao, Jenny Lee, and Sharon Smolinski for contributions to transformation development, Ambarish Nag for systems biology technical support, Stefanie Van Wychen, Corinne Feehan, and Matthew Quinn for support with biomass analysis, Ryan Sestric for technical support in the early strain selection work, and Nicholas Sweeney for supporting the continued controlled growth and harvesting of UTEX395.

REFERENCES

1. Ramli, U.S., Salas, J.J., Quant, P.A., and Harwood, J.L. (2005) Metabolic control analysis reveals an important role for diacylglycerol acyltransferase in olive but not in oil palm lipid accumulation. *FEBS J*, **272**(22), 5764–5770.
2. Nye, C., Kim, J., Kalhan, S.C., and Hanson, R.W. (2008) Reassessing triglyceride synthesis in adipose tissue. *Trends Endocrinol Metab*, **19**(10), 356–361.
3. Chaturvedi, R., Uppalapati, S.R., Alamsjah, M.A., and Fujita, Y. (2004) Isolation of quizalofop-resistant mutants of *Nannochloropsis oculata* (Eustigmatophyceae)

with high eicosapentaenoic acid following N-methyl-N-nitrosourea-induced random mutagenesis. *J Appl Phycol*, **16**(2), 135–144.

4. Chaturvedi, R. and Fujita, Y. (2006) Isolation of enhanced eicosapentaenoic acid producing mutants of *Nannochloropsis oculata* ST-6 using ethyl methane sulfonate induced mutagenesis techniques and their characterization at mRNA transcript level. *Phycol Res*, **54**(3), 208–219.

5. Doan, T.T.Y. and Obbard, J.P. (2011) Enhanced lipid production in *Nannochloropsis* sp. using fluorescence-activated cell sorting. *GCB*, **3**(3), 264–270.

6. Roessler, P.G. and Ohlrogge, J.B. (1993) Cloning and characterization of the gene that encodes acetyl-coenzyme-a carboxylase in the alga *Cyclotella cryptica*. *J Biol Chem*, **268**(26), 19254–19259.

7. Moellering, E.R. and Benning, C. (2010) RNA interference silencing of a major lipid droplet protein affects lipid droplet size in *Chlamydomonas reinhardtii*. *Eukaryot Cell*, **9**(1), 97–106.

8. Work, V.H., et al. (2010) Increased lipid accumulation in the *Chlamydomonas reinhardtii* sta7-10 starchless isoamylase mutant and increased carbohydrate synthesis in complemented strains. *Eukaryot Cell*, **9**(8), 1251–1261.

9. Davis, R., Aden, A., and Pienkos, P.T. (2011) Techno-economic analysis of autotrophic microalgae for fuel production. *Appl Energy*, **88**(10), 3524–3531.

10. Greenwell, H.C., Laurens, L.M., Shields, R.J., Lovitt, R.W., and Flynn, K.J. (2010) Placing microalgae on the biofuels priority list: a review of the technological challenges. *J R Soc Interface*, **7**(46), 703–726.

11. Bigogno, C., Khozin-Goldberg, I., Boussiba, S., Vonshak, A., and Cohen, Z. (2002) Lipid and fatty acid composition of the green oleaginous alga *Parietochloris incisa*, the richest plant source of arachidonic acid. *Phytochemistry*, **60**(5), 497–503.

12. Christie, W. (2005) *Lipid Analysis: Isolation, Separation, Identification and Structural Analysis of Lipids*. American Oil Chemists Society, 3rd Edn.

13. Griffiths, M.J. and Harrison, S.T.L. (2009) Lipid productivity as a key characteristic for choosing algal species for biodiesel production. *J Appl Phycol*, **21**(5), 493–507.

14. Bigelow, N.W., et al. (2011) A comprehensive GC-MS sub-microscale assay for fatty acids and its applications. *J Am Oil Chem Soc*, **88**(9), 1329–1338.

15. Laurens, L.M.L., Quinn, M., Van Wychen, S., Templeton, D.T., and Wolfrum, E. (2012) Accurate and reliable quantification of total microalgal fuel potential as fatty acid methyl esters by in situ transesterification. *Anal Bioanal Chem*, **403**(1), 167–178.

16. Eltgroth, M.L., Watwood, R.L., and Wolfe, G.V. (2005) Production and cellular localization of neutral long-chain lipids in the haptophyte algae *Isochrysis galena and Emiliana huxleyi. J Phycol*, **41**, 1000–1009.

17. Elliot, L., et al. (2011) Microalgae as a feedstock for the production of biofuels: microalgal biochemistry, analytical tools, and targeted bioprospecting. In Liong M. (ed.), *Bioprocess Sciences and Technology*. Nova Science Publishers, Hauppauge, NY.

18. Laurens, L.M.L. and Wolfrum, E.J. (2011) Feasibility of spectroscopic characterization of algal lipids: chemometric correlation of NIR and FTIR spectra with exogenous lipids in algal biomass. *Bioenergy Res*, **4**, 22–35.

19. Guarnieri, M.T., et al. (2011) Examination of triacylglycerol biosynthetic pathways via *de novo* transcriptomic and proteomic analyses in an unsequenced microalga. *PLoS ONE*, **6**(10), e25851.

20. Zerbino, D.R. (2010) Using the Velvet *de novo* assembler for short-read sequencing technologies. *Curr Protoc Bioinformatics* Chapter 11:Unit 11 15.

21. Schulz, M. and Zerbino, D. (2010) Oases: *De novo* transcriptome assembly for very short reads.

22. Gillham, N.W., Boynton, J.E., and Hauser, C.R. (1994) Translational regulation of gene expression in chloroplasts and mitochondria. *Annu Rev Genet*, **28**, 71–93.

23. Gruissem, W. and Tonkyn, J.C. (1993) Control mechanisms of plastid gene-expression. *Crit Rev Plant Sci*, **12**(1-2), 19–55.

24. Kirk, M.M. and Kirk, D.L. (1985) Translational regulation of protein synthesis, in response to light, at a critical stage of *Volvox* development. *Cell*, **41**(2), 419–428.

25. Mayfield, S.P., Yohn, C.B., Cohen, A., and Danon, A. (1995) Regulation of chloro-plast gene-expression. *Annu Rev Plant Physiol*, **46**, 147–166.

26. Poulsen, N. and Kroger, N. (2005) A new molecular tool for transgenic diatoms: control of mRNA and protein biosynthesis by an inducible promoter-terminator cassette. *FEBS J*, **272**(13), 3413–3423.

27. Rochaix, J.D. (1996) Post-transcriptional regulation of chloroplast gene expression in *Chlamydomonas reinhardtii*. *Plant Mol Biol*, **32**(1-2), 327–341.

28. Perkins, D.N., Pappin, D.J., Creasy, D.M., and Cottrell, J.S. (1999) Probability-based protein identification by searching sequence databases using mass spectrometry data. *Electrophoresis*, **20**, 3551–3567.

29. Jako, C., et al. (2001) Seed-specific over-expression of an *Arabidopsis* cDNA encod-ing a diacylglycerol acyltransferase enhances seed oil content and seed weight. *Plant Physiol*, **126**(2), 861–874.

30. Andrianov, V., et al. (2010) Tobacco as a production platform for biofuel: overex-pression of *Arabidopsis* DGAT and LEC2 genes increases accumulation and shifts the composition of lipids in green biomass. *Plant Biotechnol J*, **8**(3), 277–287.

31. Roesler, K., Shintani, D., Savage, L., Boddupalli, S., and Ohlrogge, J. (1997) Target-ing of the *Arabidopsis* homomeric acetyl-coenzyme A carboxylase to plastids of rapeseeds. *Plant Physiol*, **113**(1), 75–81.

32. Niu, Y.F., et al. (2011) A new inducible expression system in a transformed green alga, *Chlorella vulgaris*. *Genet Mol Res*, **10**(4), 3427–3434.

33. Chow, K.C. and Tung, W.L. (1999) Electrotransformation of *Chlorella vulgaris*. *Plant Cell Rep*, **18**(9), 778–780.

34. Jung, H., Kim, G.D., and Choi, T.J. (2006) Activity of early gene promoters from a Korean *Chlorella* virus isolate in transformed chlorella algae. *J Microbiol Biotech-nol*, **16**, 852–960.

35. Kim, D.H., et al. (2002) Stable integration and functional expression of flounder growth hormone gene in transformed microalga, *Chlorella ellipsoidea*. *Mar Bio-technol (NY)*, **4**(1), 63–73.

36. Dawson, H.N., Burlingame, R., and Cannons, A.C. (1997) Stable transformation of *Chlorella*: rescue of nitrate reductase-deficient mutants with the nitrate reductase gene. *Curr Microbiol*, **35**(6), 356–362.

37. Chen, Y., Wang, Y., Sun, Y., Zhang, L., and Li, W. (2001) Highly efficient expression of rabbit neutrophil peptide-1 gene in *Chlorella ellipsoidea* cells. *Curr Genet*, **39**(5-6), 365–370.

38. Huang, C.C., et al. (2006) Expression of mercuric reductase from *Bacillus megaterium* MB1 in eukaryotic microalga *Chlorella* sp. DT: an approach for mercury phytoremediation. *Appl Microbiol Biotechnol*, **72**(1), 197–205.

39. Wang, C., Wang, T., Su, Q., and Gao, X. (2007) Transient expression of the GUS gene in a unicellular marine green alga, *Chlorella* sp. MACC/C95, via electroporation. *Biotechnol Bioprocess Eng*, **12**, 180–183.

40. Park, H.P. and Choi, T.J. (2004) Application of a promoter isolated from *Chlorella* virus in *Chlorella* transformation system. *Plant Pathol J*, **20**, 158–163.

41. Mitra, A. and Higgins, D.W. (1994) The *Chlorella* virus adenine methyltransferase gene promoter is a strong promoter in plants. *Plant Mol Biol*, **26**(1), 85–93.

42. Franklin, S., Ngo, B., Efuet, E., and Mayfield, S.P. (2002) Development of a GFP reporter gene for *Chlamydomonas reinhardtii* chloroplast. *Plant J*, **30**(6), 733–744.

43. Mikami, K., et al. (2011) Transient transformation of red algal cells: breakthrough toward genetic transformation of marine crop *Porphyra species* (InTech) pp. 241–258.

44. Zaslavskaia, L.A., et al. (2000) Transformation of the diatom *Phaeodactylum tricornutum* (Bacillariophyceae) with a variety of selectable marker and reporter genes. *J Phycol*, **36**, 379–386.

45. Coll, J. (2006) Review: methodologies for transferring DNA into eukaryotic microalgae. *Span J Agric Res*, **4**, 316–330.

46. Atkinson, A.W., Jr., Gunning, B.E.S., and John, P.C.L. (1972) Sporopollenin in the cell wall of *Chlorella* and other algae: ultrastructure, chemistry, and incorporation of 14C-Acetate, studied in synchrounous cultures. *Planta (Berl)*, **107**, 1–32.

47. Honjoh, K., et al. (2003) Preparation of protoplasts from *Chlorella vulgaris* K-73122 and cell wall regeneration of protoplasts from *C. vulgaris* K-73122 and C-27. *J Fac Agric Kyushu Univ*, **47**, 257–266.

48. Gietz, R.D. and Schiestl, R.H. (2007) High-efficiency yeast transformation using the LiAc/SS carrier DNA/PEG method. *Nat Protoc*, **2**(1), 31–34.

49. Kindle, K.L. (1990) High-frequency nuclear transformation of *Chlamydomonas reinhardtii*. *Proc Natl Acad Sci U S A*, **87**(3), 1228–1232.

50. Dunahay, T.G., Jarvis, E.E., and Roessler, P.G. (1995) Genetic transformation of the diatoms *Cyclotella cryptica* and *Navicula saprophila*. *J Phycol*, **31**, 1004–1012.

51. Radakovits, R., Jinkerson, R.E., Fuerstenberg, S.I., Tae, H., Settlage, R.E., et al. (2012) Draft genome sequence and genetic transformation of the oleaginous alga *Nannochloropis gaditana*. *Nat Commun*, **3**, 686.

52. Hawkins, R.L. and Nakamura, M. (1999) Expression of human growth hormone by the eukaryotic alga, *Chlorella*. *Curr Microbiol*, **38**(6), 335–341.

53. Wang, P., et al. (2004) Rapid isolation and functional analysis of promoter sequences of the nitrate reductase gene from *Chlorella ellipsoidea*. *J Appl Phycol*, **16**, 11–16.

54. Seibert, P.D., et al. (1995) An improved PCR method for walking in uncloned genomic DNA. *Nucleic Acids Res*, **23**, 1087–1088.

55. Jefferson, R.A., Burgess, S.M., and Hirsh, D. (1986) beta-Glucuronidase from *Escherichia coli* as a gene-fusion marker. *Proc Natl Acad Sci U S A*, **83**(22), 8447–8451.

56. Scott, S.A., et al. (2010) Biodiesel from algae: challenges and prospects. *Curr Opin Biotechnol*, **21**(3), 277–286.

57. Wijffels, R.H. and Barbosa, M.J. (2010) An outlook on microalgal biofuels. *Science*, **329**(5993), 796–799.

58. Knoshaug, E.P. and Darzins, A. (2011) Algal biofuels: the process. *Chem Eng Prog*, **107**, 37–47.

59. Brennan, L. and Owende, P. (2010) Biofuels from microalgae—a review of technologies for production, processing, and extractions of biofuels and co-products. *Renew Sustain Energy Rev*, **14**, 557–577.

60. Kim, Y.H., et al. (2011) Ionic liquid-mediated extraction of lipids from algal biomass. *Bioresour Technol*, **109**, 312–315.

61. Siddiquiee, M.N. and Rohani, S. (2011) Lipid extraction and biodiesel production from municipal sewage sludges: a review. *Renew Sustain Energy Rev*, **15**, 1067–1072.

62. Pienkos, P.T. and Darzins, A. (2009) The promise and challenges of microalgal-derived biofuels. *Biofuels Bioprod Bioref*, **3**, 431–440.

63. Lardon, L., Helias, A., Sialve, B., Steyer, J.P., and Bernard, O. (2009) Life-cycle assessment of biodiesel production from microalgae. *Environ Sci Technol*, **43**(17), 6475–6481.

64. Shen, Y., et al (2009) Effect of nitrogen and extraction method on algae lipid yield. *Int J Agric Biol Eng*, **2**, 51–57.

65. Lee, J.Y., et al. (2010) Comparison of several methods for effective lipid extraction from microalgae. *Bioresour Technol*, **101**, S75–S77.

66. Chisti, Y. (2008) Response to Reijinders: do biofuels from microalgae beat biofuels from terrestrial plants? *Trends Biotechnol*, **26**, 351–352.

67. Takeda, H. (1993) Chemical-composition of cell walls as a taxonomical marker. *J Plant Res*, **106**(1083), 195–200.

68. Popper, Z.A. and Tuohy, M.G. (2010) Beyond the green: understanding the evolutionary puzzle of plant and algal cell walls. *Plant Physiol*, **153**(2), 373–383.

69. Takeda, H. and Hirokawa, T. (1984) Studies on cell-wall of *Chlorella*: comparison of the cell wall chemical compositions in strains of *Chlorella ellipsoidea*. *Plant Cell Physiol*, **25**(2), 287–295.

70. Walter, J.K. and Aach, H.G. (1987) Isolation and characterization of the enzymes involved in disintegration of the cell wall of *Chlorella fusca*. *Physiol Plant*, **70**, 485–490.

71. Takeda, H. (1988) Classification of *Chlorella* strains by cell-wall sugar composition. *Phytochemistry*, **27**(12), 3823–3826.

72. Rosenberg, J.N., Oyler, G.A., Wilkinson, L., and Betenbaugh, M.J. (2008) A green light for engineered algae: redirecting metabolism to fuel a biotechnology revolution. *Curr Opin Biotechnol*, **19**(5), 430–436.

9

MEIOTIC RECOMBINATION-BASED GENOME SHUFFLING OF *SACCHAROMYCES CEREVISIAE* AND *SCHEFFEROMYCES STIPTIS* FOR INCREASED INHIBITOR TOLERANCE TO LIGNOCELLULOSIC SUBSTRATE TOXICITY

DOMINIC PINEL AND VINCENT J.J. MARTIN

9.0 INTRODUCTION

With uncertainty in energy security, the rising demand and price for oil and gas derived energy, and climate change becoming omnipresent issues in society, producing alternative cleaner energy has become an important goal for governments, industry, and academia alike (1). In this climate, opportunities exist for using lignocellulosic substrates as cleaner and renewable source of sugars for bioderived products such as fuels and chemicals via fermentation. The proposed merits of bioderived fuels and chemicals from lignocellulosic substrates include reducing atmospheric carbon output, diminishing reliance on imports, and adding value to existing agricultural and forestry industries. Using waste residues as fermentable sources of sugar alleviates some concerns that arise in existing starch-based biofuel production processes. Starch-based ethanol, for example, which uses food crops as feedstock, raises concerns about

Engineering Complex Phenotypes in Industrial Strains, First Edition. Edited by Ranjan Patnaik.
© 2013 John Wiley & Sons, Inc. Published 2013 by John Wiley & Sons, Inc.

detrimentally affecting food supplies, and has at best a marginal to neutral carbon footprint, while currently relying heavily on governmental subsidies for industry viability (2). Waste streams from the forestry industry exist as by-products of an existing process, and are therefore desirable as fermentation feedstocks to add value and make full use of existing resources.

Generally, two barriers exist in the fermentation of lignocellulose-derived sugars. First, using plant biomass as a biofuel substrate is technically challenging due to its recalcitrant and variable nature. Plant material is broken down into the three major constituents of cellulose, hemicellulose, and lignin, composed of variable average amounts: 33–51% (w/w), 19–34%, and 21–32%, respectively (3,4). Lignocellulosic bioconversion seeks to access the sugars contained in these polymers for microbial fermentation to fuels and commodity chemicals, with the most developed processes leading to ethanol production. Several pretreatment practices have been developed such as acid treatment, steam explosion, and wet oxidation (5), with the aim of separating out lignin from hemicellulose, and at least partially disrupting the crystallinity of the cellulose. When biomass is broken down through such treatments a variety of inhibitory compounds derived from lignin or the breakdown products of the polysaccharides are also released (6). Inhibitors are generally separated into the groupings of furans such as 2-furaldehyde (furfural) and 5-hydroxymethyl-2-furaldehyde (HMF), organic acids such as acetic, formic, and levulinic acids, and phenolics such as 4-hydroxybenzoic acid and vanillin (7). Other stressors accompany lignocellulosic hydrolysates, such as sulfites, high dissolved solids (osmotic pressure), wood extractives, lignosulfonates, nutrient limitations, heat, and fermentation product toxicity including ethanol (8). The synergistic effects of multiple sources of inhibition have been demonstrated (6). Furthermore, it is likely that not all sources of inhibition have been accounted for in biomass hydrolysates (9). All of these sources of inhibition combine to create a toxic environment for any microorganism that might be used for the bioconversion of the lignocellulose derived sugars. To circumvent costly detoxification of the substrate prior to fermentation, it is desirable to discover or create microbial strains that can survive and ferment these substrates despite their inhibitory effects.

Yet another major barrier to the fermentation of lignocellulosic hydrolysates is in the capacity of the microorganism to efficiently metabolize all the sugars available in the substrate (1). For example, *Saccharomyces cerevisiae* is a robust fermentation biocatalyst traditionally used in the bioethanol industry, but it cannot make use of pentose sugars like xylose, which are abundant in biomass, and will only ferment hexose sugars to ethanol. The most economically viable fermentation process would make use of a microorganism that can simultaneously ferment all sugars within the substrate to the desired product, diminishing production times and infrastructure costs, such as additional fermentors for sequential fermentation by multiple biocatalysts (1).

Both challenges, tolerance to substrate toxicity and total sugar utilization, are complex phenotypes to obtain in that their optimum expression requires modulation of multiple genes and metabolic processes. As mentioned, more inhibitor-tolerant organisms such as *S. cerevisiae* are unable to use pentose sugars, while pentose-fermenting organisms are generally less tolerant to lignocellulosic substrate toxicity (10). Therefore, generating a suitable biocatalyst to ferment lignocellulosic substrates will require either reprogramming metabolism to generate pentose fermentation capabilities or increasing stress tolerance, or likely a combination of the two. The stress responses of microorganisms have been documented through gene expression studies that show they are multigenic in nature, leading at times to the differential regulation of approximately 900 genes (11). A handful of genetic targets that would make rational strain manipulation through classical molecular biology techniques plausible, such as gene knockouts or upregulation, are difficult to pinpoint, and the desired traits may not be possible without addressing large-scale multigenic cellular responses. Furthermore, using classical random mutation for strain development makes it difficult to affect a large number of mutations in a short amount of time, based on the sequential nature of mutational addition inherent in classical strain improvement schemata (see Chapters 1 and 4). Given the apparent complexity of developing lignocellulose inhibitor-tolerant traits, and the fact that the precise genetic factors involved in tolerance are largely unknown, genome shuffling is an attractive technology for developing strains that can ferment lignocellulosic substrates effectively.

To this end, a meiotic-based genome shuffling strategy was developed for yeast strains to increase tolerance to hardwood spent sulfite liquor (HWSSL) (12,13), a by-product of the acid bisulfite pulping process that can contain up to 20 g/L xylose and 30 g/L hexoses (14). HWSSL can therefore be used as a substrate for ethanol production to add value to the sulfite pulping process and to diminish biological oxygen demand resulting from disposing of unfermented SSL. However, HWSSL contains high concentrations of inhibitors commonly found in lignocellulosic hydrolysates, such as those mentioned above (15,16). It was hypothesized that quickly generating a microbe that could overcome HWSSL toxicity would require a strain development technology that could combine several mutations into a single genome in order to reprogram multigenic stress responses and achieve this complex trait. The genome shuffling approaches that were used to address this challenge will be discussed throughout this chapter and will serve as case studies of meiotic recombination-based genome shuffling (12,13).

9.1 METHODOLOGY

The two methodologies used in this study approach the challenge of HWSSL fermentation from different perspectives. The first used the yeast *Schefferso-myces stipitis* (13) as a biocatalyst, while the second used the common fermentation yeast *S. cerevisiae* (12). *S. stipitis* was chosen because of its natural ability to ferment hexose and pentose sugars, and *S. cerevisiae* was chosen because it is currently the least sensitive microorganism to the types of stress or inhibitors found in HWSSL (17). Although *S. stipitis* is generally more sensitive to HWSSL toxicity (10,16), *S. cerevisiae* populations will also die off after prolonged exposure to HWSSL (18).

9.1.1 Meiotic Recombination-Mediated Genome Shuffling

When the biocatalysts of choice are eukaryotic, as is the case with both of the above-mentioned studies, it may be possible to manipulate the natural mating cycle of the organism to propagate genetic recombination through meiosis, between mating parental mutants. The principal theories governing meiotic recombination-based genome shuffling are the same as those for protoplast fusion-based genome shuffling formats (see Chapter 4). The rationale is to create large, diverse populations of mutant strains of a particular organism as a pool of genetic diversity for combining beneficial mutations. Here, instead of using protoplast fusion to orchestrate recombination, the natural mating cycle of an organism is used in a reiterative process. Large populations of mutants are manipulated into sexually recombining to evolve individual genomes and enhance useful traits that arise from the changing genotypes. Theoretically, the natural process by which reiterative mating effects recombination bypasses the potential instability of protoplast generation, fusion, and cellular regeneration.

In order to use genome shuffling through meiotic recombination, a shuffling methodology had to be implemented for both organisms. *S. cerevisiae* has a well-understood mating cycle. The haploid generation of *S. cerevisiae* can exist in two opposite mating types, *MATa* and *MATα*. Haploid *S. cerevisiae* strains will mate under conditions that favorably promote growth of the organism, such as growing on rich media like yeast peptone dextrose (YPD). Meiotic division and subsequent sporulation can then be carried out by transfer to a nitrogen-poor media with a nonpreferred carbon source such as potassium acetate. The asci can be disrupted by enzymatic digestion of the cell wall, followed by sonication to separate sister spores. This step is crucial for genome shuffling in that it is desirable for haploids bearing one genotype to have equal opportunity to mate with haploid strains bearing differing genotypes and thus maintain the diversity of the mating populations, promoting rare mating events that bring together synergistically beneficial mutations. This process can be repeated indefinitely to accumulate beneficial mutations while retaining the ability to backcross out any deleterious mutations (Figure 9.1). To ascertain whether the mating process could be sufficiently manipulated in order to engender sexual genetic recombination on large populations, auxotrophic *S. cerevisiae* strains with differing auxotrophies were mated population-wise and screened for loss of auxotrophy (Figure 9.2). Four strains of each mating type, which were auxotrophic for 3 out of 4 essential nutrients (including leucine, histidine, tryptophan, and uracil) due to mutations affecting single genes, were mated as depicted in Figure 9.2. The entire population for mating

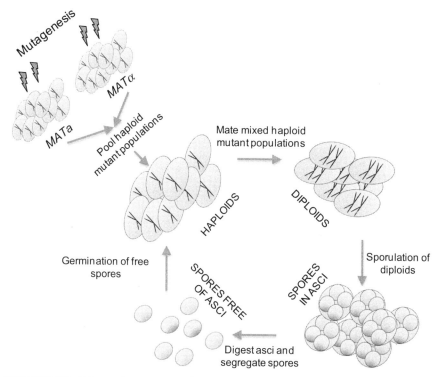

FIGURE 9.1. Schematic representation of meiotic recombination-based genome shuffling of *S. cerevisiae*. First, haploid mutant pools are generated through ultraviolet (UV) mutagenesis for each mating type (*MATa* and *MATα*). Mutant populations are mated on rich media (YPD) to obtain the diploid generation. Diploids are sporulated on potassium acetate. Spores are segregated by enzymatic cell wall degradation followed by sonication to generate a haploid generation. Haploids are germinated and mated on YPD and reiterative mating is carried out.

was initially comprised of triple auxotrophs of both mating types, based on combinations of permutations for nutritional requirements coupled with one wild-type allele. As population mating and haploid regeneration was carried out as depicted in Figure 9.2, members of the population recombined to contain more than one wild-type allele in their genome, and thus lost auxotrophy. If the wild-type alleles are treated as beneficial mutations, it is clear that they can be added together into a single genome through this process, with complete prototrophy being representative of four combined beneficial mutations. Theoretically, it was hypothesized that because the alleles corresponding to auxotrophy, or lack thereof, were on different chromosomes, after two rounds of reiterative mating two-wild type alleles could be brought together during the first mating event and four during the second, a product of mating doubly auxotrophic strains. Indeed, after the first mating round, ~35% of the

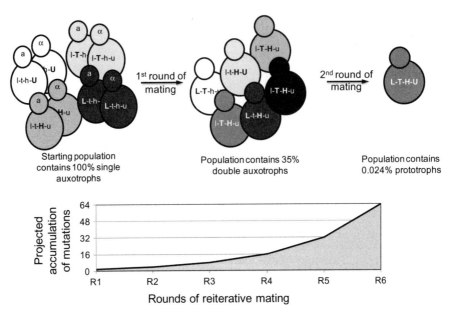

FIGURE 9.2. Testing of *S. cerevisiae* genome shuffling methodology using auxotrophic strains. A parental population of haploid strains, which were auxotrophic for 3 of 4 nutritional requirements, were reiteratively mated using meiotic recombination-based genome shuffling. The auxotrophies were based on deletion in single genes (depicted by lowercase letters: l for leucine, t for tryptophan, h for histidine, and u for uracil auxotrophies). Wild-type alleles are depicted by corresponding uppercase letters. After one round of genome shuffling, 35% of the population harbored two wild-type alleles, and after two rounds, 0.024% of the screened population showed complete prototrophy, or four wild-type alleles. The table (bottom) depicts an extrapolation of the amount of beneficial mutations (*y*-axis) that are possible through multiple rounds of meiotic genome shuffling (R1–R6).

population was comprised of double auxotrophs, while two rounds of mating led to a small percentage of completely prototrophic strains (~0.02%), which grew to nearly 1% of the population after three rounds. It should be noted that no form of population enrichment or selection for decreased auxotrophy was used between rounds, which could have led to more accelerated combination of the surrogate beneficial mutations. Also, these findings did not adhere strictly to Mendelian genetic predictions. This was likely due to the different growth rates of strains bearing differing auxotrophies (12). Because growth occurred during germination and mating, strict Mendelian statistics no longer applied. However, if one extrapolates the findings of the mating for loss of auxotrophy experiment, with a large and diverse enough population or by enriching the mating population between rounds, beneficial mutations can be accumulated indefinitely at an exponential rate as rounds of reiterative mating progress. If one contrasts this with classical strain improvement, which is sequentially subjecting a single strain to mutation for iterative improvement (see Chapter 1), as genome shuffling progresses it should greatly outpace the beneficial mutation accumulation possible in classical mutation-based strain development.

A similar mating protocol was developed for *S. stipitis*, although on a smaller scale. One of the issues working with *S. stipitis* is the insufficient information that exists on manipulating its mating cycle. When working with organisms that cannot undergo sexual recombination, or when large-scale population mating would be difficult, it is common practice to use reiterative protoplast fusion to shuffle genomes (19–25). However, Bajwa et al. (13) were successful in establishing a mating protocol for *S. stipitis*. By mating two auxotrophic strains in a similar fashion to that described in Figure 9.1, it was possible to combine two wild-type alleles and establish strains with diminished auxotrophy. Mating was made possible by spreading *S. stipitis* cultures on malt extract agar, which after incubation led to spore formation as well. Unfortunately, the percentage of the population that had combined two wild-type alleles after one round of mating was only 0.05% of the population, as opposed to the ~35% mark that was attained with *S. cerevisiae*. This low recombination efficiency suggests that future genome shuffling projects involving *S. stipitis* might benefit from protoplast fusion-based genome shuffling in order to accelerate strain evolution.

9.1.2 Inducing Genome Shuffling through Meiosis versus Protoplast Fusion

One of the goals of genome shuffling is to preserve genetic diversity within the population to promote microbial strain evolution through the combination of mutations. Maintaining large, diverse populations throughout genome shuffling is a successful method for maintaining evolution of traits of interest beyond a few rounds of reiterative mating. Additionally, the efficiency of recombination will greatly influence the outcome of the genome shuffling experiment. A lower level of recombination will require larger, perhaps prohibitively so, parental mutant populations in order to effect the beneficial combination of mutations that might comprise a small percentage of the population. Additionally, higher recombination efficiency can minimize the amount of rounds of genome shuffling required by a given project. As outlined above, with meiosis-enacted recombination, the efficiency will rely on mating and sporulation. If these steps cannot achieve a reasonable level of efficiency, protoplast fusion-based genome shuffling may be a viable option, which has a recombination efficiency that relies equally on protoplast generation, fusion, and strain regeneration efficiencies.

As genome shuffling technology is still developing, direct comparisons between the utility of differing techniques, based on separate modes of genetic recombination, are scarce. However, protoplast fusion-mediated genome shuffling has been attempted using *S. cerevisiae* as well as the meiotic recombination-mediated shuffling discussed above (Table 9.1). Shi et al. (26) have reported protoplast fusion-based genome shuffling with *S. cerevisiae* in

TABLE 9.1. Comparison of Genome Shuffling by Meiosis versus Protoplast Fusion in *S. cerevisiae*

Method	Meotic Recombinaton	Protoplast Fusion
Recombination efficiency	<35%	Unknown (100% protoplast generation, 75% cell regeneration)
Possible size of mutant mating pools	Indefinite	Indefinite
DNA level per cell	Wild-type level (haploid/diploid)	Increased per protoplast fusion
Steps involved	Mutation, mating, sporulation, spore segregation	Mutation, protoplast generation, protoplast fusion, cellular regeneration

order to engineer the trait of enhanced thermotolerance (26). In this study, a haploid *S. cerevisiae* population was made into a population of protoplasts by enzymatic digestion of the cell wall. Protoplast fusions were enacted by exposure to polyethylene glycol and fused protoplasts were regenerated on rich media containing inhibitory levels of ethanol. This study reports protoplast preparation and regeneration rates of 100% and 75%, respectively, but does not address the level of protoplast fusion attained. Assuming fusion rates are high, protoplast fusions with fungi appear to be a strong option for genome shuffling. However, one unknown to protoplast fusion-based genome shuffling is the effect that the process of reiterative protoplast fusion and cellular regeneration may have on the stability of the final strain. Shi et al. report DNA levels at 5.089, 5.144, 6.289, and 7.477 mg/g of cells for the UV mutant population and rounds one, two, and three of genome shuffling, respectively (26). It is clear that the DNA content of strains resulting from protoplast fusion increases as genome shuffling is carried out. In the end, it is yet to be determined if such an increase will affect strain stability, although after 50 generations the thermotolerant phenotype was preserved in the strains obtained through this study. As a control, classical strain improvement was carried out alongside the genome shuffling for increased thermotolerance experiment. It was found that only slight improvements to thermotolerance could be obtained by reiterative UV mutation and selection, again demonstrating the power of genome shuffling, regardless of the method of exacting recombination.

9.2 RESULTS AND DISCUSSION OF STRAIN DEVELOPMENT

9.2.1 Generation of Mutant Pools

Using the reiterative mating methodologies described above, mutant strains were genome shuffled for increased tolerance to HWSSL. UV mutagenesis was used for both *S. cerevisiae* and *S. stipitis*, with strains being exposed to UV dosages that led to a ~50% death rate. It was hypothesized that this death rate would lead to a low amount of mutations per genome, in order to minimize the chance that other deleterious mutations would mask beneficial ones. These irradiated populations were selected to enrich a population of mutants that displayed increased tolerance to HWSSL, which would be used as the parental pool for meiotic recombination-based genome shuffling.

It should be noted that other forms of mutagenesis have been applied to generate mutant pools, such as chemical mutagenesis with ethyl methane sulfate (EMS) (21). The choice of mutagen will affect the types of mutations that can occur (see Chapter 1), although most reports have used mutagens that generate point mutations, such as with UV irradiation and EMS exposure. The mutagenesis step can also be carried out on haploid (12) or diploid strains (27), or on protoplasts in the case of protoplast fusion-based genome shuffling (26). One *S. cerevisiae* genome shuffling project has reported starting with diploid strains and using a high dosage of EMS (leading to a ~90% death rate) to introduce mutation into the population (27). The rationale cited for using diploids as the initial population is that when using a higher dosage of mutagen, beneficial mutations might accompany lethiferous ones, which could mask any favorable effects. These deadly mutations will be more likely accommodated by a strain if a second copy of the wild-type genome is present. Transversely, Pinel et al. (12) chose to expose haploids of each mating type to UV mutagenesis to generate initial mutant pools (12). This measure was taken in order to ensure that if a recessive beneficial mutation is generated it will not be masked by the presence of a wild-type allele. Transversely, if a diploid is used for the parental mutant population it would be more difficult to generate recessive beneficial mutations that would be selected through screening diploid mutant populations. In this case, the diploid parental mutant population would have to be sporulated in order to screen the haploid generation and obtain strains bearing recessive beneficial mutations. Furthermore, here a lower dosage of mutagen was used in order to minimize the chances of generating deleterious mutations that would accompany the beneficial ones (12). Since most mutations will either be deleterious or have no phenotypic effect, it remains to be shown whether a high rate of mutation, introducing more than one mutation into a parental strain at once, is of benefit.

9.2.2 Screening and Selection of Mutant and Evolved Populations

Finding a large-scale, high-throughput screen is of paramount importance for any genome shuffling strategy. This may in some instances become the bottleneck for a genome shuffling project. For instance, if the phenotype of interest is increased product titers, it may be difficult to screen large populations for increased product output, especially if the product of interest requires culturing and an involved product extraction and analysis procedure. In the two studies addressed herein, growth at higher concentrations of HWSSL than the

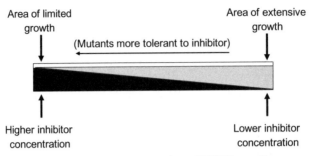

FIGURE 9.3. Profile of HWSSL gradient plate. HWSSL gradient agar plates were used for screening *S. cerevisiae* and *S. stipitis* UV mutants and genome-shuffled strains. Plastic plates ($25\,cm^2$) were elevated and HWSSL agar was allowed to solidify. The plates were then brought to a level position and overlaid with minimal media. The plates were subdivided and populations of yeast were spread for side-by-side comparison of HWSSL tolerance.

wild-type starting strains was used as a surrogate screen for tolerant organisms that could ferment HWSSL more efficiently. When the complex phenotype being addressed is increased tolerance to substrate toxicity, growth becomes an easy method for screening and therefore lends itself more easily to genome shuffling-based strain improvement. Both *S. stipitis* and *S. cerevisiae* mutants were screened primarily using gradient agar plates. The large gradient plates were made by overlaying solidified HWSSL agar with minimal media that contained similar sugar content to that found in HWSSL (Figure 9.3). This created a gradient of increasing concentration of HWSSL across the plate. The plates were divided into lanes and mutant or shuffled strains could be compared with the wild type on one plate. In this way, a large population of cells ($\sim10^7$) could be assessed for increased tolerance in any one lane of the gradient plate. The rationale behind screening large populations is to increase the probability of identifying strains that contain rare single or combined mutations.

9.2.3 Increasing HWSSL Tolerance through Genome Shuffling

At the onset of the two studies, the wild-type strains would die off readily upon exposure to undiluted HWSSL. The wild-type laboratory CEN.PK strain of *S. cerevisiae* was chosen for its general robustness and ability to mate and sporulate efficiently (28). The tolerance of the wild-type *S. cerevisiae* and *S. stipitis* starting strains to HWSSL was assessed at approximately 60% (v/v) HWSSL diluted with water. Mutant pools for reiterative mating were obtained by scraping the mutant populations from gradient plates that grew to higher levels of HWSSL concentration than the wild type. Mutants of *S. cerevisiae* grew to approximately 70% (v/v) HWSSL, and members of the final *S. stipitis* mutant population were able to grow at 75% (v/v), although they underwent three rounds of sequential UV mutagenesis as opposed to one round for *S. cerevisiae*. After five rounds of genome shuffling and selection it was found that a small number of *S. cerevisiae* genome-shuffled strains could grow and ferment undiluted HWSSL, while *S. stipitis* genome-shuffled strains could grow in 85% (v/v) HWSSL. The surrogate screen of growth on increasing concentrations of HWSSL led to strains that could ferment the sugars in HWSSL better than the wild type, for longer periods of exposure to HWSSL, seemingly indefinitely in the case of the *S. cerevisiae* strains (discussed below).

Individual strains, evolved through genome shuffling and selected on HWSSL gradient agar plates, were randomly selected from the frontier of growth in the *S. cerevisiae* study (12). Thirty mutants from the UV mutant population and 15 from rounds 1, 3, and 5 of genome shuffling with population enrichment were tested for increased tolerance to undiluted HWSSL. Although all of these strains displayed higher tolerance to HWSSL than the wild type at diluted concentrations of HWSSL agar, it remained to be shown how that tolerance would translate to exposure to 100% HWSSL liquid in shake flask fermentations. Cultures were sampled daily for viability through plate counts (CFUs/mL). The viability results showed the heterogeneity that existed within the mutant and genome-shuffled generations. For example, only 1 of the 30 tested UV mutant strains showed a noticeable increase in viability over the wild type in undiluted HWSSL. Subsequent sampling from the genome-shuffled populations led to an increased average tolerance among the 15 sampled colonies, which grew as the rounds progressed. Although heterogeneity still existed within the sampled subpopulation with regard to HWSSL tolerance at 100% HWSSL, an overall evolution toward HWSSL tolerance was achieved.

9.2.4   Tolerance to HWSSL Leads to Increased Ethanol Production

As discussed above, it should be noted that in both of the highlighted studies, growth and survivability on HWSSL were used as surrogate screens to identify better fermentative strains of HWSSL. In order to determine if increased HWSSL tolerance equates to increased ethanol productivity on HWSSL, the most HWSSL-tolerant strains were tested for sugar consumption and ethanol production at high cell densities, to mimic industrial conditions, in increased HWSSL concentrations. The three most HWSSL-tolerant *S. cerevisiae* strains were able to maintain fermentation of hexose sugars to ethanol over prolonged and repeated exposure to HWSSL (Figure 9.4) (12). Cells from the high cell density *S. cerevisiae* cultures were centrifuged and resuspended in fresh undiluted HWSSL after 2 days for the first pass and 3 days for each additional pass, up to six passes. It was shown that the HWSSL-tolerant genome-shuffled strains remained productive; that is, they survived HWSSL toxicity, consumed hexose sugars, and produced ethanol near theoretical levels of ethanol production, for all six passes, while the wild type lost the ability to ferment hexose sugars to ethanol during pass 3 in HWSSL. The *S. stipitis* study

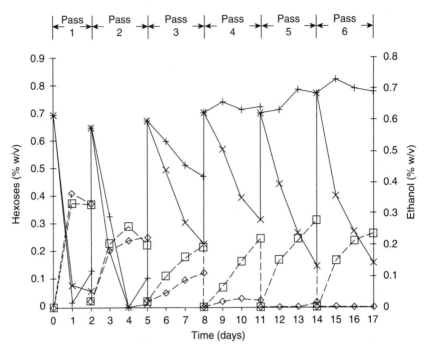

FIGURE 9.4. Ethanol production from hexose sugars in HWSSL by an *S. cerevisiae* strain obtained through genome shuffling versus wild type. The most HWSSL-tolerant strain of *S. cerevisiae* generated through genome shuffling was tested in shake flasks containing undiluted HWSSL at a high cell density for hexose sugar consumption (X sign, solid line) and ethanol production (square, dashed line) versus the wild-type parental *S. cerevisiae* strain (+ sign, solid line for sugar consumption, diamond with a dashed line for ethanol production). Cultures of each strain were resuspended in HWSSL for six passes to mimic the prolonged exposure to SSL that occurs in industrial fermentation plants.

was able to produce a strain that could produce ethanol from HWSSL glucose alone to levels between 0.15% and 0.18% (w/v) for a single 48-hour period in HWSSL, while the wild-type control was unable to ferment any of the sugars in HWSSL (13). These results suggest that using a surrogate screen of growth and viability on increased concentrations of inhibitory substrate is an appropriate screen for genome shuffling projects that aim to produce higher ethanol titers in the presence of lignocellulosic hydrolysate inhibitors.

9.2.5 Tolerance to HWSSL Leads to Cross-Tolerance to Multiple Inhibitors

Both organisms were further tested on individual inhibitors in attempts to explain the root of the tolerant phenotype. It was shown that genome-shuffled

S. stipitis strains displayed increased tolerance to the individual inhibitors like acetic acid, furfural, and HMF (13). Furthermore, HWSSL tolerance translated into cross-tolerance to three other wood hydrolysates (29). Genome-shuffled *S. cerevisiae* strains displayed tolerance to the individual inhibitors like acetic acid, HMF, hydrogen peroxide, and osmotic pressure (12). These findings demonstrate that the phenotype that has been evolved for is based on a general tolerance to multiple common sources of inhibition in lignocellulosic hydrolysates. Interestingly, the modes of tolerance to specific inhibitors do not seem to be identical. It was shown that the three most HWSSL-tolerant *S. cerevisiae* strains display increased acetic acid tolerance over the wild type only after pre-exposure to HWSSL, whereas the osmotic pressure and hydrogen peroxide tolerant traits were apparent with and without pre-exposure to HWSSL (12). This is consistent with recent findings, which suggest that tolerance to acetic acid can be an inducible response (30). Furthermore, the top-performing HWSSL strain showed increased tolerance to hydrogen peroxide, while the two other HWSSL-tolerant strains tested showed decreased hydrogen peroxide tolerance as compared with the wild type. Such findings demonstrate the heterogeneity that can exist phenotypically within a genome-shuffled population. This suggests that within a given population the ways that a single strain can arrive at the phenotype of interest are by multiple paths, in turn suggesting that the strains will harbor differing mutations and/or combinations thereof.

9.2.6 Comparison between the *S. stipitis* and *S. cerevisiae* Genome Shuffling Studies

There were differences in how the two studies were performed. A brief comparison can be seen in Table 9.2. One major difference was the size of the initial UV mutant population. The *S. stipitis* study used only six to eight individual colonies for meiotic recombination for each round of shuffling, while the *S. cerevisiae* study used the entire population that displayed more tolerance to the wild type, although the extent of the population diversity was not assessed. The tolerance to HWSSL displayed by *S. cerevisiae* grew from approximately 70% (v/v) to 100% (v/v) from UV mutant populations to round five of genome shuffling. The tolerance of *S. stipitis* increased from approximately 75% (v/v) to 85% (v/v) from UV mutant populations to round three of genome shuffling. Notably, two additional rounds of genome shuffling did not lead to a significant increase in tolerance in *S. stipitis*, while the evolution of *S. cerevisiae* progressed throughout all five rounds. This observation may stem from the fact that larger, more diverse populations were used for genome shuffling of *S. cerevisiae*, increasing the chance of bringing together synergistic mutations. Transversely, the smaller sample of tolerant individual strains used in the genome shuffling of *S. stipitis* may have exhausted the number of mutations existing in the initial mutant pool that were available for recombination, or minimized the possibility of combining rarer, synergistically beneficial

TABLE 9.2. Comparison of Genome Shuffling in *S. cerevisiae* and *S. stipitis* for HWSSL Tolerance

Organism	*S. stipitis*	*S. cerevisiae*
Mating efficiency	0.05%	<35%
Possible size of mutant mating pools	6–8 individual colonies	10^7 individual colonies
Level of HWSSL concentration tolerated after genome shuffling	85% v/v	100%
Rounds of genome shuffling that led to increased tolerance	3	5
Increased ethanol production from HWSSL	Yes	Yes
Displayed cross-tolerance to multiple individual inhibitors	Yes	Yes

mutations. Alternatively, it is possible that *S. cerevisiae* is naturally more genetically predisposed to inhibitor tolerance. Finally, the poor mating efficiency of *S. stipitis*, as shown through the mating for loss of auxotrophy assessment, may have played a role in the stagnation of population evolution. Similarly, to assess the usefulness of mating a limited number of tolerant individuals to overcome HWSSL toxicity, following round three of *S. cerevisiae* genome shuffling, five individual colonies that were able to grow at 85% (v/v) HWSSL were inter-mated using the methodology described above. However, no noticeable increase in tolerance, by HWSSL gradient plate comparisons, was witnessed (unpublished data). This finding, coupled with the lack of evolution displayed through reiterative mating past three rounds in the *S. stipitis* study, supports the theory that continued evolution of a phenotype of interest through genome shuffling is correlated to the size and diversity of the mating populations. Transversely, it was also shown by Pinel et al. (12) that using a large nonenriched population as a mating pool may act to slow the advance toward a strain displaying a desired trait as well. It was shown that if the shuffled populations are enriched for tolerant subpopulations between rounds, prior to the subsequent round of population mating, a more inhibitor-tolerant phenotype could be obtained with fewer rounds of genome shuffling. Specifically, two rounds of genome shuffling with enrichment between rounds led to populations that were more tolerant to HWSSL than four rounds without enrichment in between, shown through gradient plate screening (12). Enrichment was carried out by selecting the portion of each genome-shuffled population that displayed more tolerance than the wild type to HWSSL for use in the subsequent population-wise mating step. Here, it was hypothesized that by limiting the amount of strains displaying wild-type-level tolerance to HWSSL, the chances of combining the genomes of strains harboring beneficial mutations could be enhanced.

Although the tolerance attained by *S. cerevisiae* outpaced that attained by *S. stipitis* under the conditions described, it is important to note that *S. stipitis* has the added advantage of being able to utilize pentose sugars found in HWSSL (16). However, in undiluted HWSSL the most tolerant *S. stipitis* strains isolated in the Bajwa et al. study were unable to use the xylose present (29). Ultimately, the adverse effects of HWSSL toxicity on pentose fermentation with *S. stipitis* may prove to be a formidable barrier to overcome. However, further genome shuffling with more diverse populations could perhaps yield *S. stipitis* strains that can tolerate undiluted HWSSL and still make use of the pentose sugars found therein. To make full use of the HWSSL substrate, the ability to ferment pentose sugars would have to be engineered into *S. cerevisiae* for xylose utilization in particular, which has been shown to be a viable option (14). The tolerant strains produced in the Pinel et al. (12) study could act as background strains for rationally engineering a pentose-fermenting, HWSSL-tolerant *S. cerevisiae* strain, and further genome shuffled for strain optimization. In this way, rational strain engineering and genome shuffling may be complementary technologies for evolving complex phenotypes.

9.3 CONCLUSIONS AND FUTURE DIRECTIONS

The two studies focused on in this chapter were successful in implementing meiotic-based genome shuffling to evolve the complex phenotype of HWSSL tolerance. Juxtaposing the two cases highlights important factors that need to be addressed when developing a genome shuffling strain improvement regime. The choice of a parent organism is the first integral step. If a strain has a desirable native ability, such as pentose fermentation with *S. stipitis*, this may circumvent the need for combining rational metabolic engineering with strain evolution. Likewise, *S. cerevisiae* has a long history of use as a biocatalyst under harsh fermentation conditions, which may be based on an inherent predisposition toward robust inhibitor tolerance, and can perhaps therefore be evolved more readily to high levels of tolerance. Genome shuffling of organisms with well-understood mating cycles like *S. cerevisiae* can increase the utility of genome shuffling through reiterative mating by supporting high recombination efficiencies and the subsequent ability to bring together beneficial mutations in an exponential fashion, which is only attenuated by the size and diversity of the parental mutant populations and the high-throughput screen involved in generating mating populations. When mating efficiency is low, however, protoplast fusion-based genome shuffling may be a viable option.

These studies show the utility of using genome shuffling technology to develop strains that are tolerant of lignocellulosic substrate inhibitors. As research and interest in biofuels progress, such methods may become commonplace for developing strains that are fine-tuned to fermenting specific substrates, or creating organisms that display traits of cross-tolerance to specific inhibitors or a multitude of similar lignocellulosic substrates.

Finally, as high-throughput "omics" technologies become more common and accessible and the price of sequencing continues to diminish (see Chapter 3), the genetic changes that accompany the traits that are evolved in strain development through random approaches such as genome shuffling will help to understand complex trait evolution and inform more rational approaches to strain development. By comparing the genomes of parental strains with final mutant strains at single nucleotide resolution, it will be possible to identify the mutations that have taken place, the genes that have been targeted, and subsequently, the most important genetic factors involved in a phenotype of interest.

REFERENCES

1. Lynd, L. (1996) Overview and evaluation of fuel ethanol from cellulosic biomass: technology, economics, the environment, and policy. *Annu Rev Energy Environ*, **21**, 403–465.
2. Searchinger, T., Heimlich, R., Houghton, R.A. et al. (2008) Use of US croplands for biofuels increases greenhouse gases through emissions from land-use change. *Science*, **319**, 1238–1240.
3. Ingram, L.O., Aldrich, H.C., Borges, A.C. et al. (1999) Enteric bacterial catalysts for fuel ethanol production. *Biotechnol Prog*, **15**, 855–866.
4. van Maris, A.J., Abbott, D.A., Bellissimi, E. et al. (2006) Alcoholic fermentation of carbon sources in biomass hydrolysates by Saccharomyces cerevisiae: current status. *Antonie Van Leeuwenhoek*, **90**, 391–418.
5. Sun, Y. and Cheng, J. (2002) Hydrolysis of lignocellulosic materials for ethanol production: a review. *Bioresour Technol*, **83**, 1–11.
6. Palmqvist, E. and Hahn-Hägerdal, B. (2000) Fermentation of lignocellulosic hydrolysates. I: inhibition and detoxification. *Bioresour Technol*, **74**, 17–24.
7. Almeida, J.R.M., Modig, T., Petersson, A., Hahn-Hägerdal, B., Lidén, G., and Gorwa-Grauslund, M.F. (2007) Increased tolerance and conversion of inhibitors in lignocellulosic hydrolysates by *Saccharomyces cerevisiae*. *J Chem Technol Biotechnol*, **82**, 340–349.
8. Smith, M.T., Cameron, D.R., and Duff, S.J.B. (1997) Comparison of industrial yeast strains for fermentation of spent sulphite pulping liquor fortified with wood hydrolysate. *J Ind Microbiol Biotechnol*, **18**, 18–21.
9. Larsson, S., Palmqvist, E., Hahn-Hägerdal, B. et al. (1999) The generation of fermentation inhibitors during dilute acid hydrolysis of softwood. *Enzyme Microb Technol*, **24**, 151–159.
10. Bajwa, P.K., Shireen, T., D'Aoust, F. et al. (2009) Mutants of the pentose-fermenting yeast *Pichia stipitis* with improved tolerance to inhibitors in hardwood spent sulfite liquor. *Biotechnol Bioeng*, **104**, 892–900.

11. Gasch, A.P., Spellman, P.T., Kao, C.M. et al. (2000) Genomic expression programs in the response of yeast cells to environmental changes. *Mol Biol Cell*, **11**, 4241–4257.

12. Pinel, D., D'Aoust, F., del Cardayre, S.B., Bajwa, P.K., Lee, H., and Martin, V.J. (2010) *Saccharomyces cerevisiae* genome shuffling through recursive population mating leads to improved tolerance to spent sulfite liquor. *Appl Environ Microbiol*, **77**, 4736–4743.

13. Bajwa, P.K., Pinel, D., Martin, V.J., Trevors, J.T., and Lee, H. (2010) Strain improvement of the pentose-fermenting yeast *Pichia stipitis* by genome shuffling. *J Microbiol Methods*, **81**, 179–186.

14. Helle, S.S., Murray, A., Lam, J., Cameron, D.R., and Duff, S.J. (2004) Xylose fermentation by genetically modified *Saccharomyces cerevisiae* 259ST in spent sulfite liquor. *Bioresour Technol*, **92**, 163–171.

15. Klinke, H.B., Thomsen, A.B., and Ahring, B.K. (2004) Inhibition of ethanol-producing yeast and bacteria by degradation products produced during pretreatment of biomass. *Appl Microbiol Biotechnol*, **66**, 10–26.

16. Nigam, J.N. (2001) Ethanol production from hardwood spent sulfite liquor using an adapted strain of *Pichia stipitis*. *J Ind Microbiol Biotechnol*, **26**, 145–150.

17. Olsson, L. and Hahn-Hägerdal, B. (1993) Fermentative performance of bacteria and yeasts in lignocellulose hydrolysates. *Process Biochem*, **28**, 249–257.

18. Keating, J.D., Panganiban, C., and Mansfield, S.D. (2006) Tolerance and adaptation of ethanologenic yeasts to lignocellulosic inhibitory compounds. *Biotechnol Bioeng*, **93**, 1196–1206.

19. Cao, X., Hou, L., Lu, M., Wang, C., and Zeng, B. (2009) Genome shuffling of *Zygosaccharomyces rouxii* to accelerate and enhance the flavour formation of soy sauce. *J Sci Food Agric*, **90**, 281–285.

20. Clermont, N., Lerat, S., and Beaulieu, C. (2011) Genome shuffling enhances biocontrol abilities of *Streptomyces* strains against two potato pathogens. *J Appl Microbiol*, **111**, 671–682.

21. El-Gendy, M.M. and El-Bondkly, A.M. (2011) Genome shuffling of marine derived bacterium *Nocardia sp.* ALAA 2000 for improved ayamycin production. *Antonie Van Leeuwenhoek*, **99**, 773–780.

22. Patnaik, R., Louie, S., Gavrilovic, V. et al. (2002) Genome shuffling of *Lactobacillus* for improved acid tolerance. *Nat Biotechnol*, **20**, 707–712.

23. Wang, H., Zhang, J., Wang, X., Qi, W., and Dai, Y. (2012) Genome shuffling improves production of the low-temperature alkalophilic lipase by *Acinetobacter johnsonii*. *Biotechnol Lett*, **34**, 145–151.

24. Zhang, Y., Liu, J.Z., Huang, J.S., and Mao, Z.W. (2010) Genome shuffling of *Propionibacterium shermanii* for improving vitamin B12 production and comparative proteome analysis. *J Biotechnol*, **148**, 139–143.

25. Zhang, Y.X., Perry, K., Vinci, V.A., Powell, K., Stemmer, W.P.C., and del Cardayre, S.B. (2002) Genome shuffling leads to rapid phenotypic improvement in bacteria. *Nature*, **415**, 644–646.

26. Shi, D.J., Wang, C.L., and Wang, K.M. (2009) Genome shuffling to improve thermotolerance, ethanol tolerance and ethanol productivity of *Saccharomyces cerevisiae*. *J Ind Microbiol Biotechnol*, **36**, 139–147.

27. Hou, L. (2009) Novel methods of genome shuffling in *Saccharomyces cerevisiae*. *Biotechnol Lett*, **31**, 671–677.

28. van Dijken, J.P., Bauer, J., Brambilla, L. et al. (2000) An interlaboratory comparison of physiological and genetic properties of four *Saccharomyces cerevisiae* strains. *Enzyme Microb Technol*, **26**, 706–714.

29. Bajwa, P.K., Phaenark, C., Grant, N. et al. (2011) Ethanol production from selected lignocellulosic hydrolysates by genome shuffled strains of *Scheffersomyces stipitis*. *Bioresour Technol*, **102**, 9965–9969.

30. Wright, J., Bellissimi, E., de Hulster, E., Wagner, A., Pronk, J.T., and van Maris, A.J. (2011) Batch and continuous culture-based selection strategies for acetic acid tolerance in xylose-fermenting *Saccharomyces cerevisiae*. *FEMS Yeast Res*, **11**, 299–306.

INDEX

Engineering Complex Phenotypes in Industrial Strains, First Edition. Edited by Ranjan Patnaik.
© 2013 John Wiley & Sons, Inc. Published 2013 by John Wiley & Sons, Inc.